University of Kansas Publications in Anthropology, 16

Prehistoric Agriculture in the Central Plains

MARY J. ADAIR

Lawrence, Kansas
The Department of Anthropology
University of Kansas
1988

Designed by James Helyar
Printed by the University of Kansas Printing Service

ISBN 0-938332-14-7
ISSN 0085-2457

Contents

FIGURES

TABLES

Series Note

The Great Plains of North America are one of the world's major food growing regions. Maize, a Central American crop, wheat, a Near Eastern crop, and various sorghums, of West African origin, are grown where once prairie grasses waved. These crops were brought onto the Plains after having been domesticated by generations of selective farming and, more recently, genetic experimentation.

The food crops that were domesticated in the Great Plains almost a millenium ago are not common among food crops grown here today. Mary Adair has shown how plants native to the region and still found growing in fence rows and untilled valleys such as the common sunflower (*Helianthus annuus*) and the marsh elder (*Iva annua*), along with imports from the Southwest and the Mississippi Valley such as maize, beans, squash and the bottle gourd, were part of the "origins of agriculture" story in the Central Plains. They, along with other native plants, deserve greater understanding for their viability and possible usefulness in the face of a modern agriculture that is searching for greater adaptability in the face of excessive production inputs (fertilizers, fossil fuels, machinery), erratic markets, and vulnerability to disease of genetically specialized plants.

We welcome this addition by Mary Adair to University of Kansas Publications in Anthropology and hope it stimulates interest in food issues, historic and prehistoric archaeology, and leads to additional creative research on the domestication of foods and their cultural settings.

JOHN JANZEN, Series Editor
Department of Anthropology
University of Kansas

Preface

In some respects, the Central Plains continue to suffer from an identity crisis. Unjustly labeled the "Great American Desert" in the 1800's, they are often viewed as a flat, treeless expanse that was once home to groups of warlike nomadic Indians who traversed the endless prairies in their pursuit of the bison. In reality however, the Central Plains exhibit a considerable amount of diversity—environmentally, ecologically, and culturally. This diversity becomes readily apparent as one travels through the area today. Rolling hills, many of them loess capped, well watered gallery forests, and an abundance of flora and fauna characterize the eastern two-thirds of the area, while the western High Plains are home to strong enduring plants and a variety of animal species. It was in this diverse environmental setting that prehistoric populations adapted, eventually developing permanent villages and a marked dependence on agricultural crops.

A subsistence base including agricultural foods was well recognized among the aboriginal populations in the early historic period. Additionally, excavations of late prehistoric villages demonstrated the presence of such foods as corn, beans, and squash in a precontact setting. What was unclear however, was how long these food crops had been available to prehistoric Plains populations and whether native plants were domesticated and grown alongside the tropical cultigens. This study was undertaken to address these issues. Drawing from published and unpublished sources, the main focus of this study was to organize data on cultivated or domesticated plants and provide specific information such as varietal names, sizes. and quantification. These data could then serve as a foundation for future research as well as a basis from which to model subsistence change in the Central Plains. As new data are sure to revise our understanding of subsistence change, this study sought to emphasize the diversity in the Central Plains, particularly in regard to such factors as precipitation and level of sedentism and how such factors may have influenced the acceptance of agricultural foods.

This study was originally completed as a dissertation in 1984 for the Department of Anthropology at the University of Kansas. Since that date, portions of the text have been revised according to the availability of new data. For example, the archaeobotanical remains from several Glenwood locality sites were not available for identification or measurement in 1984 but have since become accessible, with the data incoporated into this revised version. Additionally, reconstruction formulas for *Iva annua* achenes and kernels have recently been revised with the new reconstructed sizes presented in this volume.

As a dissertation, this study merited from the advice and direction of my committee: Dr. Alfred E. Johnson, Dr. Anta Montet-White, and Dr. David W. Frayer. Subsequent to its completion, it was reviewed by Dr. Richard I. Ford, Dr. Richard A. Yarnell and Frances King. Each of the reviewers offered valuable suggestions and comments and the work towards a revision for publication was begun. During this revision stage, several individuals contributed time, advice or technical skills. Laura Poracsky of the Institute for Public Policy and Business Research at the University of Kansas redrafted the figures in Chapter 3. Carol Kuhn of the Biological Survey illustrated the front cover which depicts both native and tropical plants. Gesine Janzen prepared an earlier draft of the front cover. Joni Randel of the Department of Anthropology spent many hours typing the manuscript into the ATMS computer system for publication. James Helyar, Curator in Graphics at the Kenneth Spencer Research Library, designed the book. As editor, Dr. John Janzen spent many hours organizing these various activities so that the study could proceed to a successful completion. While I have benefited from the work and advice of all of these people, any errors in the text remain solely my responsibility.

CHAPTER 1

Introduction

A subsistence economy developed and perpetuated by a human group forms the primary interface between the particular group and their environment (Earle 1980). In addition, many aspects of the culture, such as the division of labor, social organization, technology, and length of occupation in one area, are influenced by the subsistence realm. In this regard, a change in the subsistence strategy is viewed as a major change in the cultural adaptation of a specific group. In particular, the development of agriculture is regarded as producing major evolutionary changes, a viewpoint that has always been prevalent in anthropological theory and one that has recently produced innumerable studies and research.

This is a study of subsistence change, one which involved the beginnings of agriculture in the Central Plains. The rise of food production in other areas of North America, particularly the Southwest and Eastern Woodlands, has received considerable attention. In these areas research has continued to document the various plant species cultivated or domesticated while also debating the earliest appearance of these species in an agricultural setting. Equivalent studies have not been conducted within the Central Plains, as research has neither focused on the appearance of domesticated plants nor the cultural processes involved in the change from foraging to farming communities. That such a large geographical area has been overlooked is somewhat puzzling since at least some researchers (cf Galinat 1985) believe that the Central Plains may have been geographically and culturally critical in the diffusion of some tropical cultigens out of the Southwest. Perhaps one explanation for this poor understanding of Central Plains subsistence change is that agriculture did not originate in this area but was adopted sometime after its earliest levels were clearly established in both the Southwest and Eastern Woodlands. However, the cultural, ecological, cognitive and demographic factors involved in this adoption were as important in the Central Plains as any other location in the process of a major subsistence change. The purpose of this study is to present the known data base on domesticated plants in a toxonomic framework that may aid future research in two ways. First, by compiling the evidence on cultigens, additional analyses could focus on the recovery and identification of specific species associated with critical culture-historical periods. Second, the continually changing status of domesticated plants in Central Plains archaeological contexts would only enhance a better understanding of culture change, specifically the isolation of factors involved in the change from hunting-gathering to food production.

The study area includes a major portion of the Central Plains as defined by Wedel (1961) but since the entire area is not included, perhaps the distinction "subarea" is an appropriate suffix. The limits of the study area were defined on cultural rather than geographical bases and include those lands occupied by prehistoric populations associated with Plains Archaic, Middle Woodland (Kansas City Hopewell), Plains Woodland, and Central Plains Traditions. Geographically, the area is restricted to approximately present day Kansas, northwest Missouri, and Nebraska south of the Platte River (Fig. 1.1). When applicable, adjacent areas are included in discussions since the author is fully aware of the fact that the geographical boundaries of the study area do not coincide with the entire ranges of the cultural groups included.

There are advantages as well as disadvantages in selecting the study area of this geographical extent. On the positive side, the region is large enough to include a significant number of cultigens in the study. As discussed below, while some areas within the study region have received greater attention than others in terms of excavation, data recovery, and analysis, they unfortunately have not produced sufficient amounts of domesticated plant remains on which to base a study of Central Plains subsistence. Ideally, a single drainage or locality would be best suited for a model of changing subsistence, particularly because such a model could be designed around a well developed culture-historical sequence. Since this cannot be accomplished with the current data base, the model presented in this study is generalized into a regional perspective. Such an approach is not entirely disadvantageous, however, given the current state of archaeobotanical recovery procedures discussed below. A methodological approach that is regional in scope may be the most appropriate for studying processes of cultural change and evolution (Binford 1964: 425). Once a model for the region has been proposed, individual areas can be sampled according to specific research perspectives concerning subsistence change.

A temporal limit is also appropriate for this study. Since the focus is on the development and importance of prehistoric agriculture, cultigens that appear after A.D. 1500 within the study region are not included. For periods prior to contact, six cultigens are currently recognized in an archaeological context. Included are corn (*Zea mays*), squash (*Cucurbita pepo*), bottle gourd (*Lagenaria siceraria*), marshelder (*Iva annua* var. *macrocarpa*), and sunflower (*Helianthus annuus* var. *macrocarpus*). While other tropical cultigens, such as tobacco (*Nicotina rustica*), are documented in the archaeological record from areas in proximity to this study, none have been discovered in this region. In westcentral Illinois native annuals such as goosefoot (*Chenopodium berlandieri*), ragweed (*Ambrosia trifida*), maygrass (*Phalaris caroliniana*, knotweed (*Polygonum erectum*) and little barley (*Hordeum pusillum*) have been recovered in quantity from sites and cultivation or domestication of these species has been effectively argued (Asch and Asch 1980, 1982, 1985). At present, data from the Central Plains do not support a similar development. Additionally, the protohistoric period documents the presence of watermelon (*Citrullus lanatus*) (Blake 1981) and peas (*Pisum savitum*) (Cutler and Blake 1973b) in the study area, and there are no data to suggest that these may have arrived earlier.

Considerable attention is given in the literature to the definition of terms used to describe the rather close interrelationship prehistoric populations developed with many plant species. Terms such as "encouraged," "protected," "tolerated," "propogated," "gathered," "altered," "cultivated," "domesticated," or "extended in range or habitat" are applicable in describing man-plant relationships, but they are not synonymous. From the beginning, it must be recognized that the development of agriculture involved the existence of a continuum among these relationships and while each may be recognized conceptually and adequately defined as part of the continuum, the archaeological manifestation of the term may be diffi-

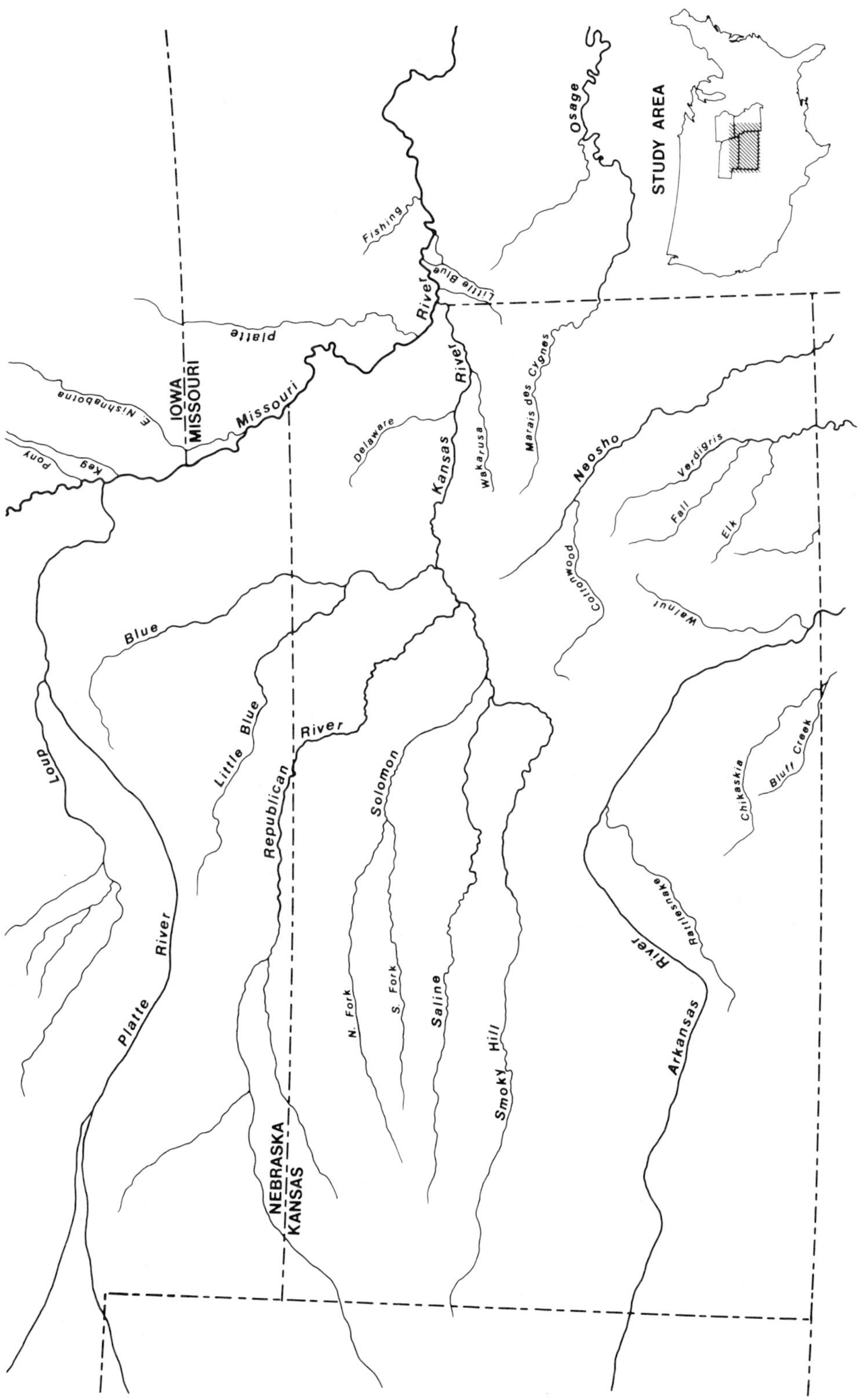

Fig. 1.1 Location of the study area.

cult, if not impossible, to identify (Asch and Asch 1982; Bronson 1977). Asch and Asch (1982) have importantly emphasized that dichotomized terms such as gardening and agriculture disguise the existence of a continuum or process while Bronson (1977:25-26) states that the use of such terms promotes an ethnocentric attitude. Bronson (1977) argues that gardening is often used to describe a technologically primitive and economically unimportant practice and agriculture is reserved for advanced economies relying primarily on domesticated plants. In reality, however, gardens can be extremely important for the economic welfare of successful modern populations (Terra 1954) and there is no reason to assume the opposite for the prehistoric past. In terms of subsistence however, the ethnographic record clearly documents varying levels of importance of cultivated crops among different populations (Denig 1961; Swanton 1946; Wilson 1977), and certainly if we accept the concept of agriculture as a developmental process or continuum, we must recognize that varying levels of importance also existed among prehistoric populations. From an evolutionary perspective, the consequences of a marked dependence on agriculture may be more important than discovering the earliest evidence for cultivation (Ford 1978).

There is no easy or standard way to deal with this problem, but a definition of terms is perhaps the most appropriate and logical place to start. Asch and Asch (1982:1) choose the term cultivation to refer to "man's deliberate encouragement of the reproduction of a species" and regard it to include the sowing of seed to the tilling of land for crops. However, this is not synonymous with agriculture, a term reserved for contexts of substantial dependence on plants grown by humans (Bronson 1977:26). While both actions may be significantly important to the economic base, a cultivator need not always be an agriculturalist. Additionally, neither case may involve domesticated plants. This biological term is strictly reserved for the specific genetic changes that sometimes accompany a plant under cultivation. In this study, the terms agriculture and domestication are preferred as they more adequately reflect our current understanding of Central Plains subsistence. Almost all plant remains discussed in this study are domesticated species and the emergence of an agricultural system is clearly recognized. The use of these terms does not ignore the processual development involved in subsistence change; as more archaeobotanical data becomes available, a stronger focus can be given to the level of cultivation.

Various theories in anthropological research have been presented as a means of explaining the initial appearance of domesticated plants, with the most recent research viewing the agricultural way of life as the result of a series of gradual changes. While new foods may have been readily absorbed into an existing subsistence practice without a noticeable effect, a complete change in the types and quantities of foods consumed was an extended process. Perhaps one reason for this gradual development was the fact that prehistoric subsistence systems were a complex interaction with a variety of cultural practices, environmental constraints, and technological capabilities. Thus a dietary change involved more than just foods, but rather was related to such factors as settlement patterns, group size and mobility, division of labor, the technological repertoire of food procurement and preparation, and concepts of what was edible, good, and necessary to the diet.

The cause(s) for such gradual changes have been expressed in many approaches, some of which rely on stress factors, such as environmental stress (Childe 1941), competition for scarce resources (Renfrew 1972), population pressure (Binford 1968b; Cohen 1977), and competitive economic strategies (Flannery 1968) as instrumental in the adoption of domesticated plants. Other theories, which promote non-stress causes, attribute the origin of agriculture to experimentation with cultigens due to increased leisure time (Sauer 1952); to optimal foraging strategies (Reidhead 1980); or to the inevitability of human nature (Braidwood 1960). Some recent theories on agricultural origins have been dominated by ecological paradigms, in which a major emphasis is placed on the interaction between human populations and their immediate environment. Demographic factors, stressed most strongly by Cohen (1977), and at one time by Binford (1968b), are viewed as causing an imbalance in a theoretical equilibrium between human populations and resources. The origin of agriculture is, in this context, the result of attempts to correct this imbalance or disequilibrium. From an archaeological perspective, therefore, the stress or disequilibrium factors should precede the first appearance of cultigens and it should be possible to argue from both a general and theoretical level that the specific factors and agriculture could have been connected as a cause and effect (Binford 1968a). At this point, it is not possible to argue from either perspective that the origins of agriculture were caused solely by a break in the equilibrium. Domesticated plants appear in the archaeological record in certain locations in the United States in preceramic contexts thousands of years prior to the recognition of significant population densities and fully developed agricultural systems. Data from the Southwest and Eastern Woodlands clearly demonstrate that certain plants preceded a sedentary lifestyle usually associated with agriculture (Binford 1983), but were added to an existing hunting and gathering mode (Ford 1981). Additionally, it is difficult to conceive on a theoretical level, a situation in which populations were *forced* to domesticate plants (i.e. Cohen 1977:195) when major changes in subsistence affect so many other aspects of culture. Furthermore, high population densities in specific localities have apparently occurred many times throughout prehistory but did not result in the adoption of agriculture (Bronson 1977:39-40).

The relationship between population densities and available resources is not denied in this study. In fact, this situation is viewed as *one* factor that importantly influenced the beginnings of agricultural systems in the Central Plains. From a theoretical perspective however, two points of caution need to be addressed here. First, there is a continuum between gathering and agriculture, and second, subsistence is an interrelated cultural aspect. Recent trends in anthropological theory concerning man-plant relationships have been advanced from an evolutionary perspective (Kehoe 1978; Rindos 1980, 1984) and are in general agreement with the ecologically oriented models which view the beginnings of agriculture as uneventful and subtle. However, these trends provide the critical basis from which to argue against unicausal (i.e. population growth) origins for agriculture while at the same time reducing the significance of invention and diffusion as necessary or appropriate factors in the development of subsistence. A basic premise of these newer models resides with the almost universal tendency of humans to exploit the biological qualities of plants to ensure the survival of human populations. The difference between conceptually distinct practices (scheduling, tending, cultivating, etc.) is a matter of degree rather than kind in man-plant relationships (Kehoe 1978:204). In this regard, Rindos (1980, 1984) has developed a model that views agriculture as an outgrowth of these evolutionary potentials which may develop whenever human populations consistently feed upon any set of plant foods. Domestication is a certain level of plant-animal interaction which favors the evolution of particular morphological changes in the plants, in addition to maintaining the actual existence of the plant, while benefiting the animal. Such an interaction, termed coevolution, is importantly not restricted to man alone (Rindos 1980: 753-754) although for prehistoric populations the primary benefit derived was the food that the plant provided. The interdependence between plants and humans is described by Rindos (1980:753) as a symbiosis in which both are positively affected. In such a situation, the subsistence strategy linked to this symbiosis is termed a domesticatory system. Conceptually, such a system may involve tending, protecting, or gardening in which domesticated plants are grown and maintained but do not constitute a major food source.

Such a positive symbiotic relationship does not however explain why domesticated plants were initially propagated intentionally by prehistoric populations, and this may very well be a situation in which a definitive answer cannot be provided. If one accepts the basic universal relationship between humans and plants as an inevitable part of a population's survival, then the exploitation of certain plants could have easily led to a domesticatory system at any point in time. From an adaptive perspective however, hunting and gathering is considered a far more successful, reliable, and conservative behavior than agriculture (Lee 1972), a factor which necessitates an explanation for the initial adoption of cultigens.

Binford (1983:204-208) addresses a similar dilemma in a discussion of mobility; his insights appear to be extremely applicable here. Contemporary hunter-gatherers do not view mobility as a bad situation in the sense that as resources are depleted in one area, the group is forced to move to another. Rather, mobility is linked to social interaction among members of the group, to critical economic information concerning the amount of food in surrounding areas, and to the group's reassurance of their economic stability. ". . . Moves were not made because they did *not* have food, but because they *did* have food" (Binford 1983:204). In other words, the mobility helps them maintain an adaptive pattern that has proven to be successful. As a corollary to Binford's quote, it is suggested that domesticated plants were initially added not as a change in prehistoric subsistence strategies but as a retention of the status quo. Domesticated plants could have easily been accepted into a hunting and gathering economic system if they were compatible with the existing and successful patterns. In particular, these plants could have provided added economic security while requiring few, if any, adjustments in a population's adaptation. This implies that the plants were not secondary or less desirable resources since their acceptance was based on selection and not necessity. In such a situation where certain plants were protected (either by spatial isolation from wild progenitors or protection from other animals), planted, and encouraged by humans, natural and/or cultural selection affected the symbiotic relationship.

Factors limiting the success of a particular domesticate are largely environmental variables, such as competition with other plants for available space and time to mature and reproduce, which are external to the symbiotic relationships between human populations and plants (Rindos 1980). Agricultural systems, which are an outgrowth of domestication, are characterized by integrated sets of activities which affect the environment inhabited by the plant. "The major effect of agriculture is to increase the carrying capacity of the environment for the domesticated plant, and this increase in plants and thus plant productivity, has major effects on human populations" (Rindos 1980:758). Plants domesticated in Archaic periods in the Eastern Woodlands would not have naturally increased their growing areas if simply left alone. A result of corn domestication left the plant unable to disperse its own seeds, hence it became dependent upon humans for this activity. *Iva* would likewise not have maintained a domesticatory status if its wild progenitors were not spatially isolated and if populations did not favor the selection of the larger seeds for harvest. In this regard, agriculture is a level or type of behavior in which domestic crops are viewed by humans to be important for their survival and well being. Subsequently environmental constraints are removed and the carrying capacity for the domesticated plants is increased. An increase in the availability of a resource is accompanied by a corresponding increase in its utilization and yield. This availability and efficiency in productivity interact in a positive feedback manner which enhances the agricultural system, as proposed by Flannery (1968).

Some element of advantage however, must be recognized as a factor initiating this behavior, since a symbiotic relationship of human populations with domesticated plants was maintained for some time prior to the development of agricultural systems. Agriculture is not an inevitable outgrowth of domestication but is an alternative restricted to human populations. In addition, agricultural systems are conceptualized as fields rather than gardens, indicating that the highly successful mobility was at some point changed to a sedentary lifestyle in which the success of the crops and economic security depended on human's constant attention at critical points. Many factors may have been involved in this change in prehistory. The relationship between population, sedentism, storage, and available resources; the genetic "fitness" of the cultivated plants; human's awareness of the potential economic stability offered by plants; and compatible environmental parameters are viewed in this study as important factors in the Central Plains.

Eventual population growth is viewed as a likely consequence of appropriate conditions. However, this growth, or reproductive potential, is not constant and is therefore difficult to measure from an archaeological perspective. Aside from the cultural practices documented as checks against overpopulation, several biological factors related to reproduction, observed in contemporary groups, offer insights into factors responsible for population growth in prehistory (Howell 1976). Research has suggested that population growth is related to the amount of group mobility (Dumond 1975; Lee 1972); the nutritional fitness of the group and particularly that of the women (Santly and Rose 1979; Stini 1975); the level of fat contained in a woman's body in proportion to her height (Frisch 1978; Frisch and McArthur 1974; Howell 1976); the total consumption of carbohydrates which increases the fat level (Kolata 1974); and the women's role in the economic system (Lee 1972; Sussman 1972). Mobility, low fat levels in women, low carbohydrate diets, and demanding roles for women in maintaining economic stability among hunter-gatherers helped keep population levels relatively low by lowering fecundity. With sedentism, however, many of these controls were significantly relaxed (Binford and Chasko 1976), indicating a greater potential for population increase. According to the archaeological record of the Eastern Woodlands, an increase in population and sedentism along with an increased consumption of carbohydrates was accompanied by chronic nutritional deficiencies and biological stress (Buikstra 1977; Cassidy 1980). While population size apparently increased steadily through time and especially after sedentism, research suggests that this trend may have been curtailed by a corresponding rise in infant mortality.

Another critical variable is significant to this argument; population density. This factor may have little to do with population growth but rather with the location and permanency of group residence. In addition, if the location of major group habitations coincides with the location of major and desired resources, any pressure caused by this localization may be economic, not demographic (Bronson 1977). The consequences are the same however; the demand has simply outrun the ability of the local habitat to maintain the commodity in steady supply, and the resulting situation may be termed "pseudo population pressure" (Bronson 1977). If domesticates are indeed food resources, then the scarcity creates a subsistence instability between plants and human populations which may be lessened by the increasing importance of domesticatory systems. Importance may be interpreted as the contribution of the behavior associated with this system to the overall subsistence of the society (Rindos 1980:758).

A combination of population growth and locational constraint would also be significant to the argument, although it may be difficult to distinguish between the two from an archaeological perspective. These two factors together however may be critical to understanding Central Plains subsistence patterns. Large sedentary populations were mainly restricted to the river valley zones where wild resources were plentiful and varied enough to sustain such a group size. Domesticated plants were also localized in these areas since rainfall amounts are inadequate for upland farming. This factor is particularly relevant for areas west of the Osage and Dissected Till Plains of eastern Kansas and will be discussed in greater detail in following chapters.

Since domesticated plants were a primary source of carbohydrates for prehistoric populations and may have been a determinant in fecundity levels, the appearance and increased availability of these foods have important implications for population growth studies. When combined with mortuary data, as demonstrated from the Eastern Woodlands, the inferences are even more pronounced. In the Central Plains, good comprehensive data on skeletal remains, mortuary practices, and population levels are lacking, while the recovery of plant remains has resulted in a sufficient data base for discussions concerning agriculture. These latter data have remained largely unpublished and unorganized, hence their potential in establishing the basis for subsistence studies has not been recognized. This study places a major emphasis on these archaeobotanical remains while available, but limited, information on population size, level of sedentism, and environment are incorporated as supportive documentation concerning the development of agriculture.

The archaeobotanical remains presented in this study are the result of both fortuitous and systematic recovery procedures. Many sites in the Central Plains included in this study were excavated two to five decades ago, before flotation or water screening were included as part of archaeological recovery procedures. However, since many of these early investigations focused on Middle Ceramic or protohistoric period occupations, charred plant remains, especially corn, were frequently recovered from intra-mural storage pits. These remains were collected by hand during excavations and little attempt was made to quantify them; a presence/absence designation was considered sufficient. Following Streuver's (1968) introduction of a more quantity control recovery technique, Central Plains archaeology focused on the recovery of botanical remains as necessary elements in the reconstruction of settlement-subsistence. Unfortunately, sampling continues to vary with individual projects and institutions and long term federal funding in some areas has resulted in a differential understanding of the regional culture-historical sequence.

The archaeobotanical data presented in this study were obtained from a variety of sources. First, the literature was searched for any reference to identified plant remains. Published and unpublished theses and dissertations, articles, cultural resource management reports, seminar manuscripts, and field notes were included as literature. Second, the actual charred remains were located from curated collections when the written reference did not include information on the sizes of the achenes or on the quantity of various identified cultigens. Unfortunately, not all curated collections could be easily located since many have been separated from other artifacts from the site. Third, the Museum of Anthropology at the University of Kansas presently curates unsorted flotation samples from certain sites. Selected samples were sorted if 1) the excavator indicated the matrix to be rich in charcoal; 2) if the site was of the appropriate time, type, or location to yield cultigens; or 3) if cultigens had been found from culturally related sites and additional remains were needed to substantiate a level of importance. Additional curated collections, such as the Schultz collection, were investigated for botanical remains since the field notes associated with this donated collection indicated the presence of cultigens.

Establishing a level of importance of cultigens was viewed as a critical factor in demonstrating the presence of agricultural systems in the Central Plains. Given the unsystematic recovery procedures, this was a difficult task since calculated quantities could not be presented. Therefore, several synergistic measures were employed to calculate the importance of each cultigen. The type or variety of each cultigen was first determined, a factor that may distinguish the level of human selection in a native plant or the productivity of a tropical cultigen within the local environment. Indigenous species were determined to be domesticated if their achene or seed size could be demonstrated to be significantly larger than their present day wild counterparts (Yarnell 1972; Ford 1981; Asch and Asch 1980). With tropical cultigens, especially maize, certain varieties are known historically to be better adapted to a prairie/plains environment than others. An assumption was made that if a low productive variety was identified, its importance as a reliable and prominent food source was doubtful. In addition, the more homogeneous the environmentally adapted variety appeared (measured in terms of phenotypic expression), the more importance it was given (Richard Ford, personal communication).

While some of the factors of recovery and identification appear to cause major biases in determining levels of importance and development of agriculture, their significance can be diminished when the amount of data and the entire study region are taken into consideration. As mentioned earlier, these biases in part dictated the selection of the study area, and while future excavations and research perspectives will undoubtedly enhance our understanding of the development of agriculture, currently available data are sufficient to provide the basis for a reasonable understanding of this process. There is no doubt that future research will add tremendously to a better understanding of Central Plains agriculture. In particular, the use of such techniques as the scanning electron microscope (SEM) can only expand our knowledge of the importance of certain plant species. There is still much to be learned from the archaeological deposits, both curated collections and those yet to be curated. This monograph was never intended to answer all of the questions or to address all of the issues involved in the emergence of food production. It is foremost an attempt to assess the current level of understanding in regards to the cultural and archaeobotanical issues involved in the development of agriculture in the Central Plains. With this goal it will hopefully provide some direction for future research.

There are six chapters in this monograph, beginning with the Introduction. Chapter 2 contains the necessary background information, such as descriptions of the geographical area, the culture history, and the archaeological sites from which charred plant remains and cultigens have been recovered. Special attention is paid to the preservation of floral materials, to recovery techniques employed during excavation, and to the identification of the archaeobotanical remains. The floral remains identified as cultigens are further discussed in Chapter 3. Size measurements and statistical measures used to describe these cultigens are available in the Appendix of this study. Chapter 4 explores several theories on the development of agriculture and applies them to the Central Plains ecological setting. Factors contributing to this major dietary change in the study area are explored, including the achievement of sedentism, the availability of productive crop plants, population growth and density, and social and technological change. In Chapter 5, botanical data from the Two Deer site (14BU55) in southcentral Kansas are presented as a case study. It is not the presence of identified cultigens at this site that is so important but rather the spatial distribution of these remains. Recovery strategies should be designed to identify and complement discard patterns and data from Two Deer suggest that such patterns may not always be related to the location of storage pits or hearths. Chapter 6 summarizes and concludes the study.

CHAPTER 2

The Cultural and Environmental Setting

In an attempt to familiarize the reader with the prehistory of the study area and with the sites discussed in this study, an overview of the geographical area, the climate and environment, and the culture history is presented. Environmental factors such as precipitation levels, temperatures, frost free days, and native flora and fauna are included in the area discussion while the culture history section focuses on distinctive ecological and biological changes recognized in the different cultural periods. Such changes include settlement patterns, subsistence strategies, and population densities. Since this study is primarily concerned with changes in subsistence, and especially with the role of plant foods, particular attention is paid to the amount of flotation conducted at sites in each cultural period and to the identification of these remains. The culture history overview presented here is not intended to summarize the entire Central Plains prehistory. In particular, several preceramic complexes are only briefly mentioned, if at all. This omission should not be taken to indicate a lesser interest or recognition in the importance of these complexes; rather the intent is to provide background information on the cultural complexes that have yielded data critical to a reconstruction of agricultural subsistence. Therefore, greatest attention is given to the Early Ceramic and Middle Ceramic culture period.

Environmental Descriptions

Physiographic regularities relevant to the settlement and subsistence activities of prehistoric populations have remained relatively unchanged over the past 2000-3000 years and are still visible in the study area today. In terms of the major physical divisions of the United States, the study area lies primarily within the Interior Plains (Schoewe 1949). Peripheral to the study area, the extreme southeast corner of Kansas and part of the Missouri River trench in the Big Bend region of central Missouri lie within the Interior Highlands and constitute sections of the Ozark Plateau province. Within the Interior Plains, the eastern one-third of Kansas and all of northwestern Missouri and southwest Iowa are included within the Central Lowland province. Additionally, Schoewe (1949:275-276) includes the area surrounding the Arkansas River trench, known as the Great Bend Lowland, within this division. The remaining western section of Kansas, western Nebraska and parts of eastern Colorado constitute the Great Plains province (Fig. 2.1).

The Dissected Till Plains were once covered by a continental ice sheet that extended beyond the Kansas River on the south and overlapped the Flint Hills on the west. Where glacial overlapping occurred, the land surface was little changed; the most common evidence of glaciation consists of a few glacial erratics scattered over the landscape. The Dissected Plains is underlain by Pennsylvanian and Permian rock formations similar to those of the Osage Plains, but the Nebraskan and Kansan glacial episodes left thick till deposits that conceal the cuesta-type topography that prevails farther south (Frye 1949; Frye and Walters 1950; Schoewe 1949).

The advance of the Kansan glacier

into northeastern Kansas scoured stream valleys and leveled uplands throughout the dissected Plains (Frye and Leonard 1952). The present day course of both the Kansas and Missouri Rivers within the study area were determined by the erosion resulting from the retreat of this glacier. During the succeeding Illinoian and Wisconsinan glaciations, wind blown sediments, or loess, were deposited over the basic cuesta-type topography. The Missouri River flows in a narrow (5 km) alluvium floodplain bordered by prominent loess capped limestone and shale bluffs and flanked in the bottoms with wooded zones. Prior to stabilization, the flat floodplain lent itself to periodic flooding and channel shifting, leaving sloughs and marshes along the old course. Streams and creeks tributary to the Missouri in the study area are usually flowing in short, deeply incised and heavily wooded ravines (Wedel 1959). Away from the Missouri River valley, erosion of the loess-till mantel has resulted in gently rolling uplands interspersed with alluvial valleys.

The Osage Plains are a large area lying south of the Kansas River and includes the subprovinces of the Cherokee Lowland of southeastern Kansas, the Chautauqua Hills south of the Neosho River, the Osage Cuestas of eastern Kansas, and the Flint Hills Upland which forms the western border of the section (Schoewe 1949). The latter two areas are of particular concern for this study. The Osage Cuesta is characterized by a series of northeast-southwest trending cuestas or ridges. The bedrock in this area consists of a series of alternating hard and soft layers of limestone and shale of Pennsylvanian age with erosion of these causing the basic cuesta type topography. The more resistant limestone strata form the highland east-facing scarp of each cuesta, while the thicker and softer shales erode more easily to form the intervening lowlands (Self 1978). The Osage, Neosho, and Verdigris Rivers flow in a general east and southeast direction (Fig. 1.1), transverse to the direction of the escarpments and against the dip of the rock formations (Schoewe 1949). These rivers and their numerous tributaries flow in wide, flat-floored valleys with well developed terraces and alluvial bottoms. Because the Osage Plains are the most generously watered area of the state, the stream banks and adjacent valley floors are heavily timbered with various hardwoods.

The Flint Hills Upland constitutes a large north-south topographic unit that extends the length of the state and whose limits are defined by the outcrop of flint-bearing Permian strata (Schoewe 1949). The characteristic east-facing escarpments are most rugged along the eastern border; to the west the unit consists of gently rolling upland surfaces often littered with fragments and nodules of flint or chert. Streams within the Flint Hills have deep channels, usually lined with outcropping rock ledges and beds consisting of angular flint and limestone pieces. Where the streams meet with less resistant shales, the valleys are wider and the valley slopes are much gentler. North of the Kansas River in the glaciated area of the state, the Flint Hills diminish in height (from a southern elevation of 1500-1600 ft) and consequently lose much of their distinctive character. Major streams and rivers that drain the Flint Hills include the Elk, Big Caney, Verdigris, Neosho, Walnut, Cottonwood, Marais des Cygnes (Osage), and Big and Little Blue (Fig. 1.1). The Kansas River is the only drainage that crosses the upland.

The Great Bend Lowland, which includes the Arkansas River Lowlands, is a broad, flat to rolling plain about 40-64 km wide that is traversed in its middle by

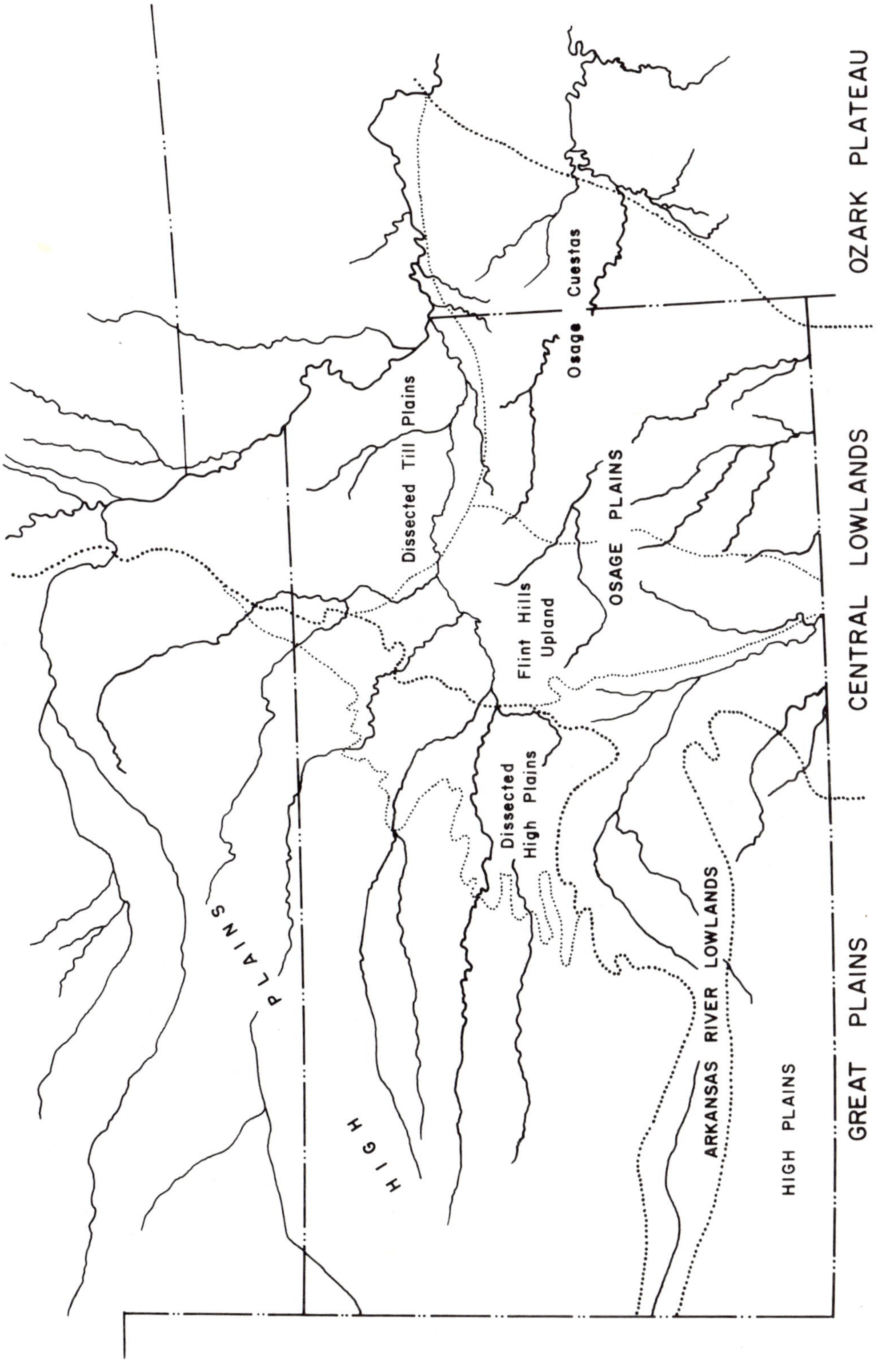

Fig. 2.1 Physiographic regions of the study area (after Schoewe 1949 and Chapman 1980).

the Arkansas River. The Arkansas River flows in a shallow channel of easily eroded Paleozoic shales and Dakota sandstone. Its low banks provide a suitable location for various grasses to flourish; the cottonwood is one of the few trees found along the river. Periodic flooding of the river has resulted in the development of several channels separated by sandy islands (Schoewe 1949). South of the river is a poorly drained area of sand dunes consisting of salt marshes, ponds, and sloughs (Wedel 1959).

The Great Plains province is the largest physiographic unit in the state and is divided by the Arkansas River into the northern Dissected High Plains and the southern High Plains. Underlain by rocks of Cretaceous age, the northern area has been irregularly dissected by stream erosion that has produced a broken landscape of high plateau-like uplands, east-facing sandstone or limestone escarpments, isolated buttes, and rolling plains (Wedel 1959). The Smoky Hills on the eastern border are carved mainly in Dakota sandstone; to the west are the Blue Hills which consist primarily of limestone and shales. Principle drainages through this area that flow in a general easterly direction include the lower Republican, Solomon, Saline, and Smoky Hill Rivers (Fig. 1.1). Their characteristic flat-floored, bluff-lined valleys are narrow (3-5 km wide) and terraced lined and support a narrow gallery forest.

Ecologically, the study area can initially be divided into the eastern tallgrass prairie and the western shortgrass plains. The hundredth meridian has often been used as a boundary line between these two areas since it coincides roughly with the line of 20 inch annual average rainfall (Wedel 1961). Dominant vegetation of the eastern tallgrass prairie includes bluestem (*Andropogon spp.*), switchgrass (*Panicum virgatum*), and Indian grass (*Sorghastrum nutans*), with various species of hickory (*Carya spp.*) and oak (*Quercus spp.*) inhabiting the low-lying valleys. Shortgrass plains vegetation is dominated by blue gamma (*Bouteloua gracilis*) and buffalo grass (*Buchloe dactyloides*) (Bare 1979). These are species that generally require less predictable moisture for growth and survival.

One of the most volatile characteristics of the prairie-plains climate is the precipitation pattern. Most of the moisture occurs as rain falling primarily from May to June, thus corresponding roughly to the spring growing season. Average annual precipitation varies from a high of 40 inches in southeastern Kansas to about 14 inches in western Kansas. Fluctuations in the moisture level are of variable duration, extent, and intensity and are relatively unpredictable. "During the last 100 years, 48 years have been below average precipitation, 44 years above, and 8 years near average" (Tomanek and Hulett 1970:203). Droughts are a common experience, although they do not always affect the entire region simultaneously or in equal measure (Wedel 1961:32). Additionally, they appear to have been major climatic events in prehistory, as the tree-ring record substantiates the presence of droughts over the last few millennia (Weakly 1943). The areas surrounding the hundredth meridian are particularly vulnerable to decreases in the mean precipitation since it demarks that line west of which agriculture is considered precarious. Its normal yearly average of 20 inches is considered almost the minimum required for successful corn growing without irrigation. The extreme variation in this rainfall line is noted from records taken in 1915 and 1936 with the former record placing the line in eastern Colorado and the latter record depicting it in eastern Kansas, a difference of about 725 miles (Wedel 1961:84).

A combination of this variable precipitation pattern and a relatively high evaporation of the moisture are reflected in the predominantly grassland biome which characterizes the study area. Evaporation, magnified by high wind velocities, is greatest in early winter, which results in very little stored moisture in the soil for the spring growing season. The native vegetation consists of shallow rooted grasses that quickly absorb the available moisture. Deciduous trees that are dependent upon stored water are usually found along stream margins, especially in the eastern section of the Central Plains (Fig. 2.2).

Temperatures in the study area exhibit a wide annual range. January temperatures are usually the lowest, averaging from 20-30°F throughout the area. The highest temperatures are experienced in July and August and normally range in the 80s for most of Kansas (Oberlander and Muller 1982: 68-70). Extremes are noted, however, with temperatures in excess of 100°F occasionally occuring throughout Kansas and usually associated with low precipitation. Temperatures nearing the 0°F are not uncommon for parts of Kansas during the winter months.

Temperature variations coincide with the length of the growing season in the study area, a factor particularly important for agriculture. In the Central Plains, about as far north as the Platte River, the period between the last spring frost and the first frost can range from 140 to 200 days (Wedel 1961; Gilbert 1980). North of the Platte River, the period ranges from 100 to 120 days.

Culture History

These physiographic and climatic features were undoubtedly an important influence on prehistoric population's adaptation to the study area. Recent studies (Rogers, 1984) suggest that the earliest adaptations may have occurred around 12,000 B.C., a time when the local environment would have been substantially different from today. The environmental conditions described in the preceding pages became prevalent at the close of the Pleistocene and correspond roughly to the beginning of the Archaic period at 10,000 B.P. The following discussion of the culture history will begin with the middle to latter part of this cultural period since little is known of previous adaptations in the study area. In addition, it is during this time that archaeologists begin to recognize for the first time a distinction between western (short-grass plains) and eastern (tall-grass prairies) adaptations, a factor that persisted until historic times (Johnson and Wood 1980:38). Table 2.1 presents a chronology of the cultural complexes discussed in this section. It is accompanied by Table 2.2, a compilation of radiocarbon dates from sites representative of the various complexes.

The arrangement of culture-historical complexes demonstrates that some areas have received greater attention than others, a factor directly related to the number and location of federally funded projects. Specific locations, such as the Kansas City and the Little Blue and Walnut River Valleys, have received adequate attention to warrant the arrangement of cultural complexes in defined temporal boundaries. In these areas, however, some complexes are listed in italics in Table 2.1, to show that while data are sufficient to indicate an occupation at a particular time, they are insufficient to define a taxonomic phase.

Archaic

Although occupation within the Central Plains is recognized during the Paleoindian and Early Archaic periods, the data are generally limited in content and extent. A clearer understanding of these early time periods is further hampered by the indication that many Paleoindian and Early Archaic sites may be deeply buried in floodplain alluvium. A date of 8060 years B.P. was recently calculated on wood charcoal recovered more than seven meters below the surface in the T-1 terrace fill of the East Fork of the Little Blue River in the Kansas City vicinity (Kopsick 1982, Schmits and Bailey 1986). Considerably more information is available for cultural complexes assigned to the Middle and Late Archaic periods. At present, research has defined six cultural complexes assigned to the Middle Archaic period and four complexes that are temporally placed within the Late Archaic period of the Central Plains.

The Middle Archaic is generally placed between 5000-2500 B.C. and the six defined phases are all located in the eastern third of the study area. Within the Little Blue Valley of eastern Kansas City, recent research has defined two Middle Archaic phases. The earliest phase, Blue Springs, is radiocarbon dated at 6580 ± 120 to 6660 ± 100 years B.P. and is characterized by small side-notched projectile points, a hunting-gathering subsistence, and temporary occupation along river terraces (Schmits and Bailey 1986) and upland ridges (Reeder 1977). The second Middle Archaic Jacomo phase is represented at both habitation sites and burials and is radiocarbon dated at 5550 ± 100 and 5590 ± 120 years B.P. (Schmits and Bailey 1986:227) and 5420 ± 210 years B.P. (Baker et al. 1985). This phase is identified by medium to large-sized expanding stemmed and corner-notched point forms, small temporary sites with a limited amount of debris, and burials without grave offerings.

The Logan Creek phase, originally defined by Kivett (1962) for the Logan Creek site (25BT3) in northeastern Nebraska, has been recognized in southcentral Nebraska, northcentral Kansas, and northeastern Kansas (Grange 1980; McDonald 1967; Witty 1982; Wright

Table 2.1. Culture-Historical Complexes of the Central Plains subarea.

Period	*Dates*	*Specific Localities*			*General Areas*		
		Kansas City	*Little Blue Valley*	*Walnut Valley*	*Eastern Kansas*	*Northcentral Ks. Southcentral Neb.*	*Northeast Ks. Southeast Neb.*
	A.D. 1500						
					Pratt	Loup River	
		Steed-Kisker	May Brook				
Middle					Smoky Hill/Bluff	Classic	
Ceramic	A.D. 1000		Lake City		Creek/Pomona	Republican	Nebraska
				Bemis Creek		Solomon Rv.	
	A.D. 800	*Late Woodland*	Woods Chapel				
							Sterns Creek
		Edwardsville		Butler	Grasshopper Fls/	Keith	
					Greenwood		
		Kansas City			Schultz/Cuesta		
			Kansas City				
			Hopewell				
Early							
Ceramic	A.D. 1	Trowbridge					
Early							
Woodland	500 B.C.	*Early Woodland*	Bowlin	Walnut	Walnut		
		Nebo Hill	Nebo Hill	El Dorado	El Dorado		
Archaic	2000 B.C.				Munkers Creek/		
			Jacomo	Chelsea	Black Vermillion		Logan Creek
	5000 B.C.		Blue Springs				

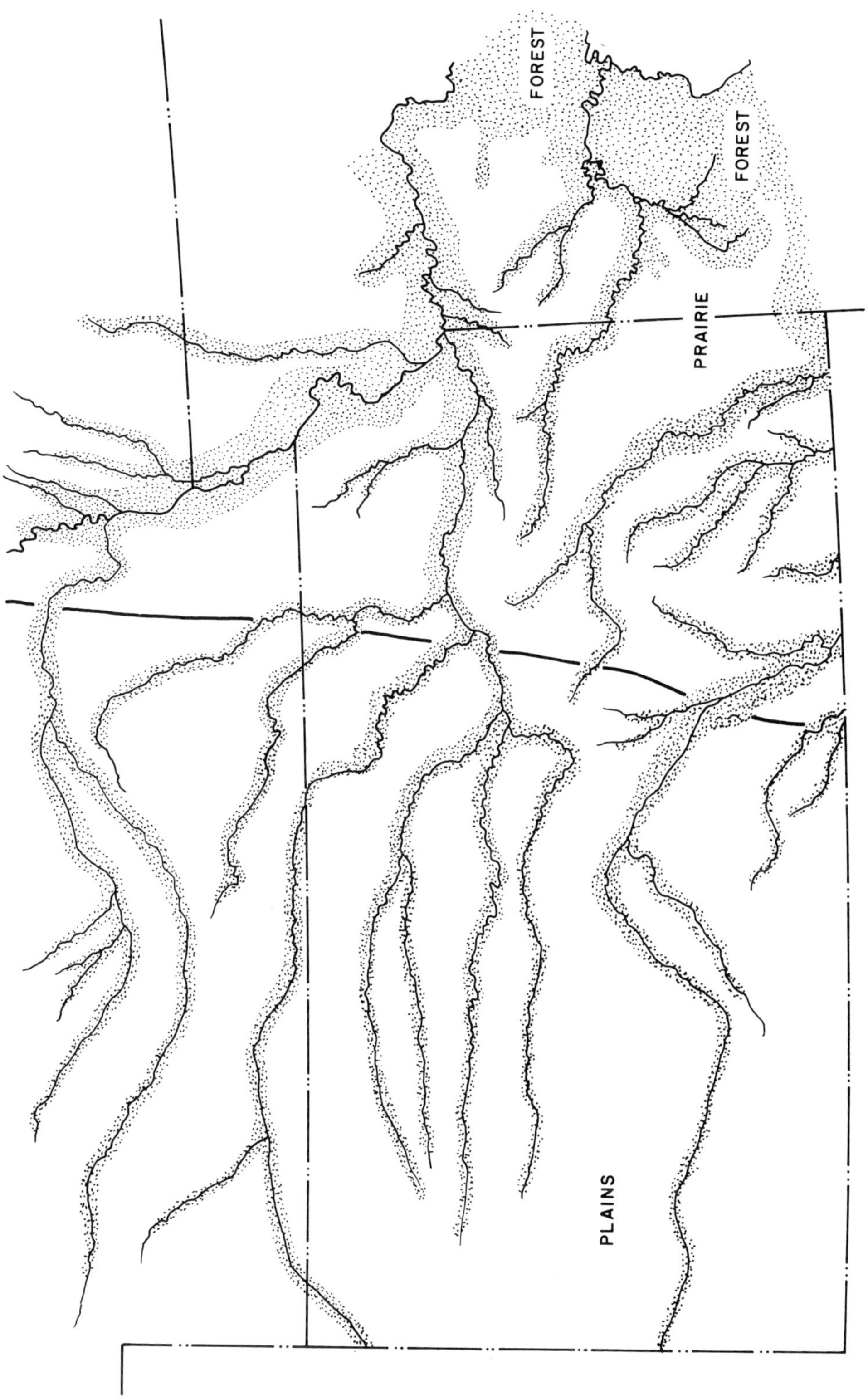

Fig. 2.2 Extent of forest, prairie and plains vegetation throughout the study area (adapted after Küchler 1964).

Table 2.2. Radiocarbon Dates for Sites Discussed in Text.

Cultural Affiliation	*Site Number*	*Name*	*Lab Number*	*Calendar Date*[a] *(uncorrected)*	*Reference*
Archaic	14PO1	Coffey III-1	WIS-774	3130 ± 65 B.C.	Schmits 1978
		III-1	WIS-776	3080 ± 65 B.C.	
		III-2	WIS-778	3120 ± 70 B.C.	
		III-3	WIS-779	3190 ± 65 B.C.	
		III-5	N-1549	2850 ± 95 B.C.	
		III-5	WIS-618	3205 ± 70 B.C.	
		III-5	WIS-623	3220 ± 70 B.C.	
		III-7	WIS-624	3290 ± 70 B.C.	
		III-7	WIS-634	3175 ± 70 B.C.	
		III-7	WIS-628	3210 ± 70 B.C.	
		III-7	UGa-382	3555 ± 105 B.C.	
		III-8	WIS-629	3335 ± 70 B.C.	
		III-8	WIS-636	3305 ± 70 B.C.	
		III-8	N-1550	3730 ± 135 B.C.	
		IIa	WIS-715	4335 ± 145 B.C.	
		IIa	WIS-711	3405 ± 70 B.C.	
Blue Springs	23JA143		Beta-8535	4710 ± 100 B.C.	Schmits, ed. 1986
			Beta-12001	4630 ± 120 B.C.	
Jacomo	23JA155	Cold Clay	Beta-8536	3640 ± 120 B.C.	Schmits, ed. 1986
			Beta-12000	3600 ± 100 B.C.	
			Beta-1325	2295 ± 170 B.C.	
			DIC-1678	2590 ± 150 B.C.	
			DIC-1679	2230 ± 95 B.C.	
Jacomo	23JA277		Beta-11684	3470 ± 210 B.C.	Baker et al. 1985
Nebo Hill	23JA35	Turner-Casey	Beta-1873	2600 ± 115 B.C.	Schmits and Wright 1982
Nebo Hill	23JA110	Sohn	DIC-913	1020 ± 490 B.C.	Reeder 1981
Nebo Hill	23CL11	Nebo Hill	UGa-1329	A.D. 1540 ± 100*	Reid 1980a
			UGa-1330	A.D. 345 ± 105*	
			UGa-1332	1605 ± 65 B.C.	
			UGa-1342	A.D. 480 ± 60*	
Nebo Hill	14MM27		TL	2000 ± 1000 B.C.	Blakeslee and Rohn 1982
			TL	1940 ± 400 B.C.	
			TL	1725 ± 570 B.C.	
			TL	1650 ± 375 B.C.	
			TL	1620 ± 300 B.C.	
			TL	1590 ± 370 B.C.	
			TL	1500 ± 350 B.C.	
			TL	1430 ± 335 B.C.	
			TL	1260 ± 300 B.C.	
Early Woodland	23JA159	Traff	UGa-2535	395 ± 70 B.C.	Wright 1980
			UGa-2405	505 ± 80 B.C.	
Early Woodland	23JA36		UGa-1873	450 ± 85 B.C.	Ziegler 1981b
Early Woodland	23JA40		UGa-2351	350 ± 100 B.C.	Ziegler 1981a
Bowlin	23JA38	Bowlin Bridge	Beta-1326	490 ± 90 B.C.	Peterson 1982
Early Woodland	14PO1	Coffey	DIC-1357	370 ± 60 B.C.	Schmits 1981
			DIC-1358	530 ± 55 B.C.	
Archaic	14BU9	Snyder	N-1279	2880 ± 105 B.C.	O'Brien 1984
			N-1280	2650 ± 125 B.C.	
			N-1278	2030 ± 110 B.C.	
			N-771	1960 ± 160 B.C.	
			N-770	1700 ± 140 B.C.	
			N-1277	1290 ± 85 B.C.	
			N-1551	220 ± 110 B.C.	
			N-1276	110 ± 80 B.C.	Grosser 1977
			N-769	20 ± 110 B.C.	
Middle Woodland Kansas City Hopewell	14WY1	Trowbridge	N-971	A.D. 130 ± 105	Johnson 1979
			N-972	A.D. 260 ± 100	
			N-970	A.D. 320 ± 100	
			N-973	A.D. 360 ± 125	
			N-974	A.D. 400 ± 105	

Table 2.2. Radiocarbon Dates for Sites Discussed in Text. (continued)

Cultural Affiliation	*Site Number*	*Name*	*Lab Number*	*Calendar Date*[a] *(uncorrected)*	*Reference*
Middle Woodland Kansas City Hopewell	23PL1	Renner	M-572	A.D. 8 ± 250	Johnson 1979
			GaK-1166	A.D. 100 ± 90	
			M-571	A.D. 108 ± 200	
			M-573	A.D. 438 ± 200	
			M-454	A.D. 687 ± 250	
Middle Woodland Kansas City Hopewell	23PL4	Young	N-1197	40 ± 110 B.C.	Johnson 1979
			CWR-180	A.D. 610 + 60-75	
Middle Woodland	23SA128	Fisher-Gabert	N-1049	A.D. 70 ± 105	Kay and Johnson 1977
			N-1050	A.D. 270 ± 105	
			N-1051	A.D. 270 ± 105	
			N-1052	A.D. 350 ± 105	
			N-1053	A.D. 260 ± 105	
			N-1054	A.D. 440 ± 105	
Middle Woodland	23JA110	Sohn	D1C-914	270 ± 195 B.C.*	Reeder 1980
			D1C-912	A.D. 150 ± 425	
			D1C-911	A.D. 230 ± 75	
Middle Woodland	23JA238	Black Belly	Beta-8537	A.D. 10 ± 90	Schmits, ed. 1986
			D1C-1680	A.D. 330 ± 45	
Middle Woodland	23JA143		D1C-1683	A.D. 330 ± 45	Schmits, ed. 1986
Middle Woodland	23JA36		UGa-1874	A.D. 430 ± 170	Ziegler 1981b
			UGa-1875	A.D. 595 ± 210	
Middle Woodland	23JA40		UGa-2350	A.D. 100 ± 140	Ziegler 1981a
Early Late Woodland	23CL199	Yeo	UGa-1449	A.D. 100 ± 100*	O'Brien 1982
			UGa-1450	A.D. 1035 ± 165*	
			UGa-1452	A.D. 930 ± 295*	
			UGa-1453	A.D. 765 ± 65	
Cuesta	14LT304		GaK-4793	A.D. 1240 ± 180*	Witty 1982
			GaK-4794	A.D. 670 ± 140	
			GaK-4795	A.D. 1830 ± 70*	
Schultz	14GE41		UGa-2142	A.D. 415 ± 85	Parks 1978
			UGa-2143	A.D. 700 ± 165	
Schultz	14GE312	Elliott	UGa-381	A.D. 15 ± 75*	O'Brien et al. 1973
Keith	25RW28	Doyle	SI-68	A.D. 520 ± 45	Grange 1980
			UM-466	A.D. 580 ± 100	
Keith	25FT18		M-841	A.D. 820 ± 200	Long and Mielke 1966
			SI-126	A.D. 370 ± 100*	
Keith	25FT70		SI-197	A.D. 690 ± 80	Long and Milke 1966
Keith	14PH4	Woodruff Ossuary	C-928	A.D. 607 ± 240	Wedel and Kivett 1956
Woods Chapel	23JA238	Black Belly	Beta-12786	A.D. 560 ± 60	Schmits, ed. 1986
			Beta-8538	A.D. 640 ± 60	
Lake City	23JA85	Sperry	UGa-1867	A.D. 730 ± 70	Brown 1981
			UGa-1868	A.D. 805 ± 60	
			UGa-1869	A.D. 695 ± 65	
Late Woodland	23CL276		UGa-2714	A.D. 730 ± 135	McHugh 1982
			Gx-6312	A.D. 830 ± 140	
			Gx-6313	A.D. 745 ± 130	
			UGa-2857	A.D. 710 ± 60	
Greenwood	14MM13	Stiles	UGa-4088	A.D. 155 ± 145*	Blakeslee and Rohn 1982
			UGa-4089	615 ± 195 B.C.*	
			UGa-4091	A.D. 725 ± 75	
Greenwood	140S347	Cow-Killer	Beta-12906	A.D. 650 ± 100	Reynolds, 1985
Butler	14BU57		UGa-3137	A.D. 1195 ± 235*	Brown 1982b
Butler	14BU31		UGa-2560	A.D. 735 ± 55	Leaf 1979
			UGa-2807	A.D. 710 ± 75	
Grasshopper Falls	14JF331	Anderson	GaK-1734	A.D. 760 ± 90	Reynolds 1979
Sterns Creek	23CC28	Walker Gilmore	M-1130	4140 B.C.*	Haas 1983
			M-1129	A.D. 920 ± 150*	
			UGa-2804	A.D. 1140 ± 55	

Table 2.2. Radiocarbon Dates for Sites Discussed in Text. (continued)

Cultural Affiliation	*Site Number*	*Name*	*Lab Number*	*Calendar Date*[a] *(uncorrected)*	*Reference*
			UGa-2803	A.D. 1285 ± 55	
			UGa-2802	A.D. 1195 ± 70	
			UGa-2801	A.D. 1195 ± 65	
			UGa-2800	A.D. 1520 ± 50*	
			UGa-2799	A.D. 960 ± 75*	
Bemis Creek	14BU55	Two Deer	UGa-2500	A.D. 885 ± 65	Adair and Brown 1981
			UGa-2501	A.D. 965 ± 45	
			UGa-2502	A.D. 1625 ± 140*	
			UGa-2503	A.D. 860 ± 55	
			UGa-2504	A.D. 1000 ± 135	
			UGa-1345	A.D. 1060 ± 60	
			UGa-1346	A.D. 980 ± 80	
Pomona	14D019	Hatcher	UGa-4705	A.D. 875 ± 65	Brown 1983
Pomona	14BO319	Roth	I-12,327	A.D. 720 ± 180	Brogan 1982
			I-12,338	A.D. 850 ± 90	
Pomona	14CF506	Winn	I-9656	A.D. 1495 ± 145	Rohn et al. 1977
			I-9657	A.D. 1370 ± 150	
Pomona	14CF508	Anderson	I-9654	A.D. 1060 ± 80	Rohn et al. 1977
			I-9655	A.D. 1245 ± 135	
May Brook	23JA115	Seven Acres	UGa-2352	A.D. 1335 ± 65	Brown 1981
			UGa-2353	A.D. 1245 ± 55	
May Brook	23JA43	Maybrook	D1C-1522	A.D. 1170 ± 90	Schmits 1982
			D1C-1526	A.D. 1220 ± 130	
			D1C-1523	A.D. 1800 ± 80*	
			D1C-1524	A.D. 530 ± 125*	
May Brook	23JA238	Black Belly	D1C-1603	A.D. 1270 ± 65	Schmits, ed. 1986
Middle Ceramic	14BU71		UGa-1347	A.D. 1210 ± 75	Fulmer 1977
Bluff Creek	14SR303	Buresh	GaK-2796	A.D. 1080 ± 90	Witty 1983
			GaK-2797	A.D. 900 ± 110	
Bluff Creek	14SR305	Nulik	GaK-2798	A.D. 1190 ± 110	Witty 1983
Smoky Hill	14PO4	Budenbender	M-869	A.D. 1190 ± 150	Johnson 1973
Smoky Hill	14GE600	Witt	UGa-825	A.D. 1355 ± 60	O'Brien 1984
			UGa-826	A.D. 1115 ± 65	
			UGa-824	A.D. 1250 ± 75	
Smoky Hill	14LC301	Root	I-509	A.D. 987 ± 100	Witty 1983
Smoky Hill	14RY401	Griffing	GaK-1446	A.D. 1630 ± 100*	O'Brien 1984
			BaK-1447	A.D. 1690 ± 70*	
			GaK-1448	A.D. 1580 ± 70*	
			GaK-1449	A.D. 1630 ± 80*	
			GaK-1450	A.D. 1700 ± 70*	
			UGa-465	A.D. 1090 ± 60	
			UGa-827	A.D. 1280 ± 60	
			UGa-828	A.D. 1140 ± 60	
Smoky Hill	14GE21	Miller	SI-230	A.D. 1030 ± 90	Long and Mielke 1966
			SI-231	A.D. 1180 ± 80	
			SI-232	A.D. 1540 ± 100*	
Smoky Hill	14CY30	Woods	M-113	A.D. 1176 ± 150	Witty 1963
			GaK-295	A.D. 1200 ± 120	
Solomon River	14JW301		GaK-593	A.D. 910 ± 100	O'Brien 1984
Solomon River	140B27	Sumter	GaK-1088	A.D. 680 ± 90*	O'Brien 1984
			GaK-1090	A.D. 570 ± 80*	
			GaK-1535	A.D. 1480 ± 80	
			GaK-1536	A.D. 590 ± 70*	
Solomon River	14ML371		GaK-1091	A.D. 830 ± 80	Lippincott 1978
Solomon River	14ML15		GaK-1089	A.D. 1090 ± 80	Lippincott 1978
			GaK-804	A.D. 610 ± 90*	
Solomon River	14ML16		GaK-638	A.D. 1190 ± 90	Lippincott 1978
Upper Republican	14CY102	Moll Creek	GaK-803	A.D. 1530 ± 90*	O'Brien 1984
Upper Republican	25FT35	Mowry Bluff	GaK-1675	A.D. 630 ± 210*	Wood 1969a
			GaK-1674	A.D. 1630 ± 90*	
			GaK-1676	A.D. 1480 ± 120	

Table 2.2. Radiocarbon Dates for Sites Discussed in Text. (continued)

Cultural Affiliation	*Site Number*	*Name*	*Lab Number*	*Calendar Date*[a] *(uncorrected)*	*Reference*
			WIS-318	A.D. 1160±55	
			WIS-319	A.D. 1180±55	
			WIS-324	A.D. 1020±60	
Nebraska	14DP10	Nuzum	WIS-326	A.D. 1090±55	Wood 1969a
			GaK-1672	A.D. 1530±90*	
			GaK-1673	A.D. 1580±90*	
Nebraska	14DP13	Squaw Creek	M-1069	A.D. 1560±75*	O'Brien 1984
Glenwood	13ML126		WIS-632	A.D. 1220±45	Tiffany 1981
			WIS-633	A.D. 965±45	
			WIS-716	A.D. 1110±60	
Glenwood	13ML128		WIS-560	A.D. 1130±50	Tiffany 1981
			WIS-565	A.D. 1160±50	
			WIS-566	A.D. 1095±55	
Glenwood	13ML129		WIS-559	A.D. 1130±55	Tiffany 1981
			WIS-562	A.D. 1185±55	
			WIS-564	A.D. 1125±60	
Glenwood	13ML130		I-6295	A.D. 1238±90	Tiffany 1981
			I-6296	A.D. 1005±100	
			I-6297	A.D. 1185±110	
Glenwood	13ML135		WIS-713	A.D. 1090±60	Tiffany 1981
			WIS-717	A.D. 975±60	
			WIS-719	A.D. 1040±60	
Glenwood	13ML136		WIS-697	A.D. 1070±60	Tiffany 1981
			WIS-698	A.D. 1175±55	
			WIS-699	A.D. 1085±55	
			WIS-702	A.D. 1260±50	
Glenwood	13ML138		WIS-729	A.D. 940±55	Tiffany 1981
Glenwood	13ML139		WIS-691	A.D. 1510±45	Tiffany 1981
			WIS-700	A.D. 1115±45	
			WIS-701	A.D. 1100±55	
Steed Kisker	23CL113	Friend and Foe	M-9026	A.D. 1100±110	Calabrese 1969
			GaK-1993	A.D. 970±90	
			GaK-1994	A.D. 1050±90	
			GaK-1995	A.D. 760±80*	
Steed Kisker	23PL13	Steed Kisker	M-1397	A.D. 860±110	O'Brien 1984
			M-1395	A.D. 1000±110	
			GaK-590	A.D. 1080±80	
			M-1395a	A.D. 1110±110	
			M-1398	A.D. 1210±100	
			M-1399	A.D. 1230±100	
Steed Kisker	23PL48	Gresham	M-1182	A.D. 875±150	O'Brien 1984
Steed Kisker	23PL6	Vandiver Mound	GaK-330	A.D. 1290±80	O'Brien 1984
Steed Kisker	23PL54	McClaron		A.D. 1260±90	Shippee 1972
Steed Kisker	23PL4	Young	M-2347	A.D. 1070±110	O'Brien 1984
			M-2346	A.D. 1290±100	
Steed Kisker	23PL16	Coons	UGa-466	A.D. 1305±60	O'Brien 1984
			UGa-467	A.D. 905±60	
			UGa-392	A.D. 1315±60	
			UGa-379	A.D. 770±110*	
Steed Kisker	23CL118	Butcher	M-2179	A.D. 780±150	O'Brien 1984
Steed Kisker	23CL109	Hulse	UGa-1445	A.D. 1085±70	O'Brien 1984
			UGa-1446	A.D. 690±90*	
			UGa-1447	A.D. 1115±75	
			UGa-1448	A.D. 1255±100	
Steed Kisker	14WY7	Calovich Mound	UGa-1454	A.D. 1105±80	Barnes 1977

* date is not in agreement with material cultural remains and/or with radiocarbon dates from components of the same complex; therefore it does not adequately reflect the prehistoric occupation.

[a] dates based on half-life of 5568 or 5570.

1982; Wood 1977). Diagnostic artifacts include small to medium sized side-notched triangular points, various scrapers, grinding stones and pigments. Subsistence focused mainly on hunting bison, deer, antelope, beaver, and wolf and collecting freshwater mussel. No structures are associated with this phase.

Sites with occupations assigned to the Munkers Creek and Black Vermillion phases are located primarily within the Flint Hills physiographic region. Dated between 3050-3550 B.C., the Munkers Creek phase is recognized at the William Young (Witty 1962, 1982), Cow Killer (Reynolds 1984), and Roniger sites. The Black Vermillion phase, similar in temporal and spatial arrangement, is identified at the Coffey site (Schmits 1976, 1978, 1981). Both phases may represent an adaptation to climatic fluctuations associated with the close of the Altithermal. The subsistence economy included hunting and gathering of various animals and plants. Sites are primarily located on terraces above major streams and appear to have been periodically occupied and perhaps re-occupied. Shallow pits, stonelined hearths, and isolated post stains suggest a more substantial occupation than before.

The Chelsea phase is restricted to the southern Flint Hills region where it is recognized at the Snyder (Grosser 1970, 1977), Nuttle (Fulmer 1976, Artz 1982), Milbourn (Root 1981a), and perhaps 14BU81 (Artz 1981) sites. Radiocarbon dates of 2880-2200 B.C. place this phase at the end of the Middle Archaic period. Settlements were restricted to river terraces where the subsistence economy focused on a variety of wild animals. Limited flotation precludes any discussion of the importance of plants although grinding stones do occur within the habitation debris.

The Late Archaic phases recognized within the study area include the Colvin phase of the Osage Cuestas (Rohn, Stein and Glover 1977), the El Dorado and Walnut phases of the Flint Hills region (Artz 1981; Grosser 1970, 1977; Padgett and Blakeslee 1982; Schmits 1981), and the Nebo Hill phase of the Dissected Plains and northern Osage Cuestas (Blakeslee and Rohn 1982; Reeder 1980; Reid 1983; Schmits and Wright 1982). Of these four phases, two large sites in the study area offer the best data for describing the Middle Late Archaic adaptation. One of the sites is Coffey (14PO1) at Tuttle Creek Lake in north-central Kansas (Schmits 1978, 1981); the other is Nebo Hill (23CL11) in northwestern Missouri (Reid 1980a, 1983). Many other Archaic sites have been investigated but generally do not offer the information on subsistence found in these two sites.

At Coffey, five buried cultural horizons in Unit III, dated from 3730 ± 135 to 2850 ± 95 B.C. (Schmits 1978), were the focus of investigations by Kansas State University and the University of Kansas from 1970 to 1975. Excavations yielded a quantity of artifacts, including chipped and ground stone tools, bone tools, cores and lithic manufacturing debris, burned earth and charcoal. Faunal remains documented the prehistoric exploitation of large and small mammals, amphibians, reptiles, birds, and fish (Schmits 1978). Each of the occupation zones contained features, primarily hearths representing either food processing loci or nuclei for surrounding activities. Feature fill was saved for flotation, yielding thousands of charred seeds. Identified plants included goosefoot (*Chenopodium spp.*), knotweed (*Polygonum sp.*), Solomon's seal (*Polygonatum sp.*), bulrush (*Scirpus sp.*), sunflower (*Helianthus sp.*) and grape (*Vitis sp.*) (Table 2.3).

Although the exploited species of flora and fauna varied between occupation levels (Schmits 1978:154-157), similar subsistence activities are represented in each of the zones. Included are butchering and processing of large game, scraping, hideworking, woodworking, and seed grinding, with the latter activity represented by more tools than the hunting activities.

The successive cultural horizons at Coffey are interpreted as short term, probably seasonal, encampments by small groups of hunters and gatherers. Similar adaptations along stream or river margins are noted for other Late Archaic sites in the study area, including Snyder (14BU9) and Milbourn (14BU25) in southcentral Kansas (Grosser 1973; Leaf 1981; Root 1981a) and Cow Killer (14OS347) and Williamson (14CF330) in eastcentral Kansas (Schmits 1980; Reynolds 1984; Witty 1982). At Coffey, the floral and faunal remains, in conjunction with the numerous tools, indicate a diffuse, broad based subsistence economy existed throughout the various occupations. The stability of such an economy is documented by the repeated occupation at the site during the latter part of the Archaic over a period of approximately 200 years.

Within the Kansas City locality, the Nebo Hill phase (Reid 1983) is recognized as the major Late Archaic adaptation. Excavations at the type site, 23CL11, identified the complex as representing a "hunting and gathering pottery-equipped population centered in the Lower Missouri Valley" (Reid 1983:34) during the second millennium B.C. Diagnostic artifacts include lanceolate points, rectangular manos, three-quarter grooved axes, bifacial hoes and gouges, and fiber-tempered ceramics. Subsistence strategies focused on the upland forest and forest-edge species with faunal remains including deer, fish, turtle, duck and squirrel. Plant remains were represented at 23CL11 by black walnut (*Juglans nigra*), goosefoot (*Chenopodium sp.*), hackberry (*Celtis sp.*), and various species of grasses (Root 1979). The Nebo Hill site was interpreted as a warm weather occupation on an unsheltered interfluve of the Missouri River. Smaller, low density sites in tributary terraces (such as Sohn, 23JA110 [Reeder 1980]) are suggested to be the winter encampments by the Nebo Hill people. In addition, Reid (1983:34) states that this settlement pattern is "indirect evidence for a seasonal cycle that included a warm (wet) weather phase of social ag-

Table 2.3. Probable Plant Foods from Archaic and Early Woodland Sites.

	Coffey 14PO1[a]	*23JA143*[g]	*Nebo Hill 23CL11*[b]	*Turner-Casey 23JA35*[c]	*23JA9*[c]	*Traff 23JA159*[d]	*23JA36*[e]	*23JA40*[f]
Chenopodium spp.	4500+	7	122			947		
Amaranthus sp.	9	16	1			83	103	
Polygonum sp.	6+		2					1
Vitis sp.	x			1				
Iva annua			1					
Helianthus spp.	10							
Euphorbia sp.								3
Celtis occidentalis	1		10					
*Juglans nigra**		29	191.6g	51.5g	x	1.2g		
Carya spp.	2	11	1.3g	32.5g	x	94.1g	100+	
Corylus americana			.4g					
Quercus spp.				16.0g				

[a] Schmits 1978, 1981 (Horizons III 1-8); [b] Root 1979; [c] Schmits, ed. 1982; [d] Adair 1980; [e] Ziegler 1981b; [f] Ziegler 1981a; [g] Schmits, ed. 1986.

x present but not quantified in original manuscript.

*weights of nutshell fragments are provided when available; otherwise counts are listed.

gregation in the loess uplands and a cold (dry) weather phase of social segmentation in the tributary valleys.''

Although the subsistence practice and settlement pattern of the Nebo Hill phase appears to be similar to other Late Archaic adaptations in eastern Kansas, several lines of evidence suggest that this phase is significantly different from those to the west. First, two diagnostic artifacts, the lanceolate points and fiber-tempered ceramics, are suggested to have served important political and social functions in addition to their use as utilitarian implements in the subsistence economy (Reid 1980a). Additionally, such prestige items that may demark a wealthy or distinctive individual are viewed as more characteristic of an expansionist clan or lineage system than of a simple band level society (Reid 1983:35).

A comparison of the morphological attributes of the bifaces from Nebo Hill to those from collections to the east, led to the suggestion that the development and recognition of this phase may be attributed to influence from eastern populations. Plains Archaic occupations throughout eastern Kansas, such as Coffey, do not exhibit this eastern orientation.

Table 2.3 presents a summary of a variety of potentially edible plant foods that have been recovered from Archaic sites in the study area. Although not represented by many sites, the information indicates that Archaic subsistence strategies involved the exploitation of one or two select plant communities or resource zones. The Late Archaic Nebo Hill occupations at 23CL11 and Turner-Casey (23JA35) indicate a concentration on upland climax and floodplain forest plant communities such as black walnut and hickory. At Coffey, the emphasis was placed on species adapted to alluvial soils or edge environments, as depicted in the quantity of goosefoot recovered. To date, no cultigens have been recovered from any Archaic occupation in the Central Plains.

Early Woodland

Following the Nebo Hill phase, further cultural developments in the Kansas City locality are recognized with the appearance of an Early Woodland occupation. Taxonomically defined as the Bowlin phase (Schmits and Bailey 1986) several sites, including Traff (23JA159) (Wright 1980), 23JA36 and 23JA40 (Ziegler 1981a, 1981b) and Bowlin Bridge (23JA38) (Peterson 1982) have been radiocarbon dated to 500-350 B.C. (Table 2.2) and have yielded artifacts stylistically similar to those recovered from Early Woodland phases in west-central Illinois (Montet-White 1968). Early Woodland ceramics and points have also been observed by the author in the private collections of local amateur archaeologists. At present, little is known of this adaptation although available data suggest that the settlement-subsistence pattern does not differ substantially from that presented for the previous Late Archaic phases (Wright 1980) (Table 2.3).

Most information on the Early Woodland period comes from Traff (23JA159), a multi-component site located on a low terrace of the Little Blue River in Jackson County. Investigations by the University of Kansas in 1977-78 yielded a few ceramic fragments; chipped stone tools, including sub-triangular unstemmed and stemmed points and bifacial tools; and floral and faunal remains. These artifacts were distributed in and around several hearth remains, leading the excavator to suggest that they represent remnants of a set of domestic activities (i.e. cooking, food preparation) that occurred in an open air setting

(Wright 1980:76). Subsistence remains included deer, bison, beaver, raccoon, goosefoot, pigweed, hickory, and black walnut (Wright 1980; Adair 1980). The latter were extracted both by hand during excavation (mainly large nutshell fragments) and by systematic collection of matrix samples from each unit and feature by flotation.

The density of cultural remains at identified Early Woodland sites suggests that the inhabitants were likely small groups of hunters and gatherers. Brown and Ziegler (1981:437) also suggest that the settlement pattern followed seasonal fluctuations since most sites are situated on the same soil association that is characterized by frequent flooding. These sites were therefore most likely occupied during the late fall through the winter, the driest period of the year.

Although similarities have been noted to Early Woodland phases in the east, artifacts from the Kansas City locality also exhibit similarities to materials from the more western Walnut phase identified at Snyder (14BU9) (Grosser 1977) and Coffey (14PO1) (Schmits 1981). Radiocarbon dates for these two sites range from 530-20 B.C. (Table 2.2), roughly the same temporal span as seen in the Kansas City locality. Triangular corner-notched projectile points recovered from the Walnut phase components at Snyder and Coffey are morphologically similar to those associated with the Early Woodland occupation at Traff (Schmits 1981:203). The exact connection between these two complexes has yet to be clearly identified; forthcoming studies will hopefully clarify the cultural affiliation between these complexes.

Early Ceramic

The Middle Woodland occupation in the Kansas City area is best known from extensive fieldwork and analyses of Kansas City Hopewell sites. Several of these sites, including Trowbridge (14WY1), Young (23PL4), Renner (23PL1), and Yeo (23CL199) are discussed in this study. A single Middle Woodland site in central Missouri, Fisher-Gabert (23SA128), is also included. Within the Little Blue Valley in Missouri, Middle Woodland occupations have been identified at Sohn (23JA110) (Reeder 1980), 23JA36 (Ziegler 1981b), Black Belly (23JA238) (Schmits 1986), and 23JA143 (Peterson and Schmits 1982). These occupations have been interpreted as indicative of several poorly known Middle Woodland complexes that may have developed from the Early Woodland Bowlin phase and differ considerably from the Kansas City Hopewell cultural complex (Schmits and Bailey 1986).

The Kansas City Hopewell variant is suggested to be a manifestation originated and perpetuated by migrations from and/or contact with Hopewell populations to the east (A. Johnson 1974). With the emergence of the Early Woodland materials in the Kansas City locality, however, this contact may have begun prior to the onset of the Middle Woodland period, perhaps as early as the Late Archaic (Reid 1983). Regardless of the origins of the Kansas City Hopewell, general trends of stylistic change through time in both ceramics and bifacial tools observed in the Illinois Valley are paralleled in the Kansas City locality, although with a temporal lag (A. Johnson 1979:90-91). Long term social interactions apparently occurred between the two areas although the mechanisms by which these were accomplished is somewhat sketchy. Initial evidence for trade and contact comes from Reid's (1976) conclusion that approximately 45 percent of the Middle Woodland bladelets from the Trowbridge site were manufactured from exotic cherts extracted from the Boone Formation limestones of central Missouri. Johnson (1979:90) postulates that this contact was facilitated by using the Missouri River as a corridor and that the Kansas City Hopewell Aker site (23PL43) may represent a regional exchange center in this scheme. On a larger scale, however, there is little to suggest that the Kansas City Hopewell populations significantly participated in an extensive exchange network or interaction sphere. Exotic artifacts, such as copper or helmet shells, are infrequent in the collections, while anthropomorphic figurines and platform pipes, although similar to styles from Illinois, are manufactured from local Kansas City clays (A. Johnson 1979).

A series of radiocarbon dates (A. Johnson 1979, 1983) places the Kansas City Hopewell period from about the time of Christ to A.D. 800. Three developmental phases of the Kansas City Hopewell, based largely on diagnostic ceramic design and decoration in association with radiocarbon dates, have recently been identified (Johnson, personal communication). The Trowbridge phase (A.D. 1-250) is the earliest and is followed by the Kansas City phase (A.D. 250-500). The final phase, representing the Early Late Woodland in the Kansas City area, is the Edwardsville phase (A.D. 500-800). Several of the large Kansas City Hopewell villages, such as Trowbridge, Renner, and Young are believed to have been occupied for the initial 500-600 years of Hopewell duration while the Yeo site is representative of the Early Late Woodland phase.

While many aspects of the Kansas City Hopewell remain unknown, the settlement-subsistence strategies of these people have received considerable attention (A. Johnson 1974, 1976; E. Johnson 1975; Adair 1977; Bell 1976; Brown 1975b; Pullen 1976; O'Brien 1982; O'Brien et al. 1979). Current understanding of these strategies indicates the Kansas City Hopewell were well adapted to the ecological milieu around the junction of the Kansas and Missouri Rivers (Fig. 2.3). Primary environmental zones exploited for resources were the floodplain forests and upland oak-hickory forests and prairies. Major faunal subsistence items included white-tailed deer, bison, elk, raccoon, turkey, and various amphibians and reptiles. Floral remains, summarized in Table 2.4, attest to the gathering of a variety of locally available plants. Both the diversity of wild plants and the appearance of domesticates is readily observable from Table 2.4 and demonstrates a marked change over the recognized exploitation patterns of the preceding Archaic and Early Woodland populations. Although nuts, particularly hickory and hazel, continue as significant food items, small seeded annuals and fruit seeds are represented in sizeable amounts. Seeds of goosefoot (*Chenopodium spp.*), pigweed (*Amaranthus sp.*) and knotweed (*Polygonum sp.*) have been recovered from almost every site where flotation has been conducted (Table 2.4). The varying amounts of fruits such as plum (*Prunus sp.*), persimmon (*Diospyros virginiana*) and pawpaw (*Asimina triloba*) may be explained by the percentage of different ecological zones exploited within the vicinity of each site. For example, Trowbridge is situated on a bluff, enabling the inhabitants to gather more upland species than available to Young, located on a low-lying terrace.

The settlement pattern consisted of three distinct site types. The first and perhaps best known are large permanent villages situated at the mouths of tributaries of the Missouri River (Young, Renner) or at the junction of small tributaries of the Kansas River (Trowbridge). Smaller, short-term camps, such as Deister and Yeo, are possibly ancillary to the villages and are located throughout

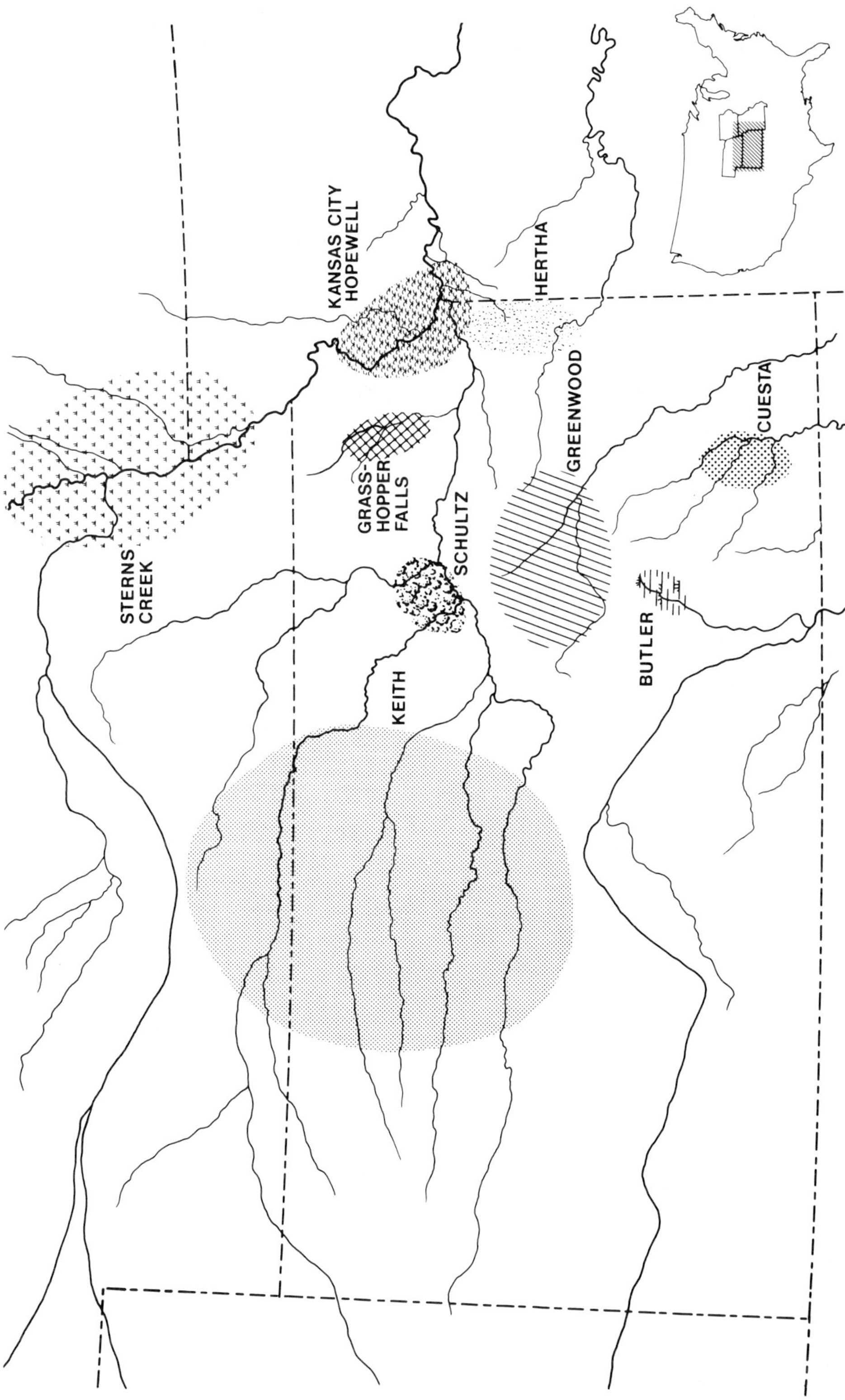

Fig. 2.3 Location of Early Ceramic complexes within the study area

the tributary valleys (A. Johnson 1976). These may have functioned as resource procurement stations or they may represent daughter communities originating from population pressure at the large villages. Burial mounds, which comprise the third site type, have largely been destroyed; consequently their association with the villages or camps is based mainly on a few sites (Larsen and O'Brien 1973; Tjaden 1974; Wedel 1943).

Several cultural and biological factors separate the Kansas City Hopewell from preceding populations. First, the large villages, the presence of numerous storage pits, and the well executed subsistence strategy provide the first evidence of sedentism in the Kansas City locality (cf Reid 1980b). Unlike the Late Archaic Nebo Hill folk, the Kansas City Hopewell were able to support their lifestyle and population by exploiting a greater number of food sources (E. Johnson 1975; Adair 1977). The establishment of large villages in major river valleys with their varied and dependable natural resource base served to reinforce the settlement-subsistence pattern of the Kansas City Hopewell population and allowed for the growth of large aggregations of people that could be maintained over long periods of time. This adaptation remained viable despite the nonexploitation of migratory waterfowl and aquatic resources, a strategy so characteristic of Middle Woodland populations to the east (Parmalee et al 1972; Braun 1977).

Population levels above those suggested for the preceding cultural periods were also a characteristic of the Middle Woodland period. Although no estimates on population have been suggested, the number of habitation sites and their accumulation of cultural debris might alone indicate a moderately dense population in the Kansas City area. Additionally, several lines of evidence suggest that a gradual increase in population occurred over the 800 year span of the Kansas City Hopewell occupation.

Table 2.4. Probable Plant Foods from Middle Woodland and Early Late Woodland Sites.

	Young 23PL4[a]	*Trowbridge* 14WY1[b]	*Fisher-Gabert* 23SA128[c]	14LT304[d]	*Yeo* 23CL199[e]
Chenopodium spp.	42		x		8
Amaranthus sp.	106	1	x		
Polygonum sp.	25		x		3
Iva annua	21	18*	2*		58*
Vitis sp.	6	6	x		
Euphorbia sp.	15		x		
Helianthus sp.	1			1	
Ambrosia trifida		22	x		
Diospyros virginiana		1			
Asimina triloba		40			
Prunus sp.	1			1	
Rubus sp.	3		x		
Cucurbita pepo		1*			
Zea mays		100*		2*	
*Juglans nigra***	3.8g	.3g			.1g
Carya spp.	4.0g	5.7g			82.8g
Corylus americana	7.0g	139.7g			.3g
Quercus spp.		1.1g			

[a] Adair 1977; [b] E. Johnson 1972; [c] Robinson 1976; [d] Tom Witty, personal communication; [e] King 1982a.

* represents a cultivated variety.

[x] present, not quantified in manuscript.

** weights of nutshell fragments are provided when available; otherwise counts are listed.

An increased expansion of the area occupied by these people is a primary response suggested to have occurred (A. Johnson 1979). A seriation of Kansas City Hopewell ceramics which display variations in form and design with radiocarbon dates (Johnson and Johnson 1975) suggest a gradual south to north expansion through time from the Kansas City area to present day St. Joseph. The later sites may possibly represent daughter communities established by fissioning off part of a larger population under demographic stress (A. Johnson 1979: 88; Brockington 1977).

The incorporation of domesticated plants into the diet is viewed as an additional response to population expansion (A. Johnson 1979). Horticulture, as a subsistence redundancy, would have increased the available food supply without altering the basic exploitation pattern of the Middle Woodland populations. This suggestion is viewed as somewhat speculative at this time, and is discussed in later sections of this study.

While floral remains have been recovered from many Kansas City Hopewell sites, the recovery techniques have been neither systematic nor consistent between sites. At Trowbridge, the matrix from nine of the recorded 45 features were floated and no midden samples were collected. Excavations at Renner were conducted prior to the introduction of flotation and most of the floral remains recovered have since disintegrated. At Young and the Middle Woodland occupation at Fisher-Gabert, the matrix from all features was floated and most of the botanical remains have been analyzed (Adair 1977; Robinson 1976). Consequently, the number of identified taxa of charred seeds, particularly from Young, is considerably larger than known from other Kansas City Hopewell sites. The Yeo site is the only limited activity Kansas City Hopewell site where flotation samples were taken; matrix from the five features was analyzed for botanical remains. No midden samples have been selected from any Kansas City Hopewell site.

Middle Woodland Kansas City Hopewell influence apparently spread out of the Kansas City locality to several locations in eastern Kansas. Data are sufficient to define two Early Ceramic phases that show association with these populations; Cuesta and Schultz complexes. The Cuesta phase was identified from sites in southeast Kansas, located in the lower Verdigris, Fall, Elk, and Neosho River valleys (Marshall 1972; Brogan 1981). The Schultz phase is less clearly defined but encompasses an area in northcentral Kansas around the Republican and Big Blue River valleys. Both phases date to the latter half of the Kansas City Hopewell sequence, or after approximately A.D. 250. Data are insufficient to determine whether these phases originated from contact with the Kansas City Hopewell or as a result of direct migration out of the Kansas City locality.

Settlement-subsistence practices of the Cuesta phase somewhat parallel those from the Kansas City area. Nucleated villages consisting of several large houses, isolated households, and small hamlets have been identified for the Cuesta phase. The thick midden accumulations at the villages suggest that these sites were occupied for an extended period. Floral and faunal remains attest to the exploitation of native plants and animals (Marshall 1972:91-92) with recent excavations providing evidence of the minimal use of charred corn (Rowlison 1980; Brogan 1981) (Table 2.4).

The Schultz complex is best known from investigations of several burial mounds (Eyman 1966; Schultz and Spaulding 1948; O'Brien 1984). The mounds are described as small structures about 20-40 ft (6-12 m) in diameter and 2-4 ft (.6-1.2 m) high. A central burial chamber is usually present but grave goods accompanying the interment are not abundant. O'Brien (1984) has concluded that the treatment is most like Hopewellian but not as elaborate. Ceramics recovered from several of these mounds exhibit Hopewellian designs such as rocker and dentate stamping (O'Brien et al. 1973) while village sites thought to be associated with these mounds have yielded plain surfaced styles commonly associated with the Early Late Woodland period (Parks 1978). Several radiocarbon dates are in agreement with this later temporal placement (Table 2.2).

In the Kansas City locality, the following Late Woodland period is believed to be partially contemporaneous with terminal Kansas City Hopewell and may owe some of its development to influence from this preceding cultural manifestation. Several Late Woodland cultural phases have recently been identified for the Little Blue Valley (Table 2.1). The earliest, Woods Chapel, is represented at Black Belly (23JA238) and is comparable in most respects to Johnson's Early Late Woodland Edwardsville phase (Schmits and Bailey 1986:234). The Lake City phase, represented at Sperry (23JA85) may also be recognized over a much wider area since diagnostic ceramics have been found in adjacent drainages. Clearly, these complexes need to be better documented so that the taxonomic designations do not become redundant. Despite these potential problems, however, the local Late Woodland phases represent marked changes in settlement-subsistence from that observed for the Kansas City Hopewell. The late Late Woodland settlement pattern recognized at Sperry differs in that no inter-site relationship such as the base camp/ancillary camp association seems to have existed. Late Woodland occupations recorded to date in the Little Blue River valley, in the uplands of Platte and Clay counties, in Brush Creek valley, and in the Fishing River drainage, are small sites consisting of meager amounts of cultural debris (Brown 1981; A. Johnson 1974; Logan 1979; Martin 1976; Schmits and Bailey 1986). Brown (1981) further suggests they represent short term occupations, probably by hunter-gatherers.

To date, only one single component Late Woodland Lake City phase site has been excavated in the Kansas City locality (Brown 1981). Extensive flotation of both feature matrix and midden fill at Sperry (23JA85) indicate a reliance on such native plant foods as goosefoot (*Chenopodium sp.*) and pigweed (*Amaranthus sp.*) but a complete absence of cultivated plants (Table 2.5). The presence of manos and other food processing tools suggests, however, that plants still contributed to the diet. Additional Late Woodland components have been investigated in the Smithville area, although the two sites reported, 23CL274 and 23CL276, are multi-component Late Woodland and Steed-Kisker sites (McHugh 1982). Domesticated marsh-elder (*Iva annua*) and corn (*Zea mays*) were recovered from several features that contained both diagnostic Late Woodland and Mississippian ceramics. Although radiocarbon dates from these features confirm a Late Woodland occupation (Table 2.2), the association of the cultigens cannot be determined. Until more Late Woodland sites are excavated and the question of horticulture during this period is resolved, empirical data are lacking to suggest that large upland Late Woodland farmsteads were established after the adoption of corn agriculture (A. Johnson 1983:197), a development described for the Illinois valley (Styles 1981). This point is discussed with greater attention in Chapter 4 of this study.

In the preceding pages, the cultural remains from components assigned to the Walnut phase were discussed as resembling materials from the Early Late Woodland complex in the Kansas City locality. It has been further suggested that the Walnut phase, rather than the Early or Middle Woodland complexes in the Kansas City area, may be ancestral to the Plains Woodland and that this latter tradition may date from the early centuries A.D. to A.D. 1000 (Johnson, personal communication). The temporal and spatial extent of the Kansas City Hopewell variant however was significant enough to perhaps also potentially influenced the development of Plains Woodland complexes. Future studies will hopefully focus on the development of the Plains Woodland as it is a major cultural tradition in the Central Plains. Radiocarbon dates from several Plains Woodland sites (Table 2.2) fall within the A.D. 700-1000 range, suggesting a general contemporaneity with both the Early Late Woodland (Edwardsville and Woods Chapel phases) and the Late Woodland (Lake City phase) developments of the Kansas City area.

In discussing additional cultural complexes in the study area, it becomes apparent that different taxonomic systems have been used in designating recognized manifestations. The literature contains complexes assigned to both the Midwest Taxonomic System (McKern 1939) and the Cultural-Historical Integration system (Willey and Phillips 1958), with the latter being the most recent and widely used. Many times however, a specific cultural complex designated as a "phase" does not properly describe the culture in the manner in which it was intended (Willey and Phillips 1959). Geographical distributions and/or temporal ranges are often more extensive than indicated by the definition of this taxonomic unit. Therefore, it has been suggested that the term variant (Lehmer 1971; Krause 1969; Johnson, in press) be adopted, since it appears to be better suited for the organization of

Table 2.5. Probable Plant Foods from Late Woodland and Plains Woodland Sites.

	Sperry 23JA85[a]	*Stiles 14MM13*[b]	*14BU57*[c]	*23JA238*[d]	*14BU32*[e]	*14BU9*[f]	*14BU31*[f]
Chenopodium spp.	36	3	3	5	2	13	344
Amaranthus sp.	5		4				10
Polygonum sp.							5
Vitis sp.			1			1	
Euphorbia spp.			12				
Helianthus sp.							1
Galium sp.			1		1	6	1
Ambrosia sp.			1				
Rubus sp.		3	1				
Vicia sp.		3					
gramineae					2	28	8
Viola sp.				13			
Portulaca sp.				8			
Celtis occidentalis						5	
Zea mays		1	1				
*Juglans nigra**		13	4	3	1.1g	3.7g	
Carya spp.	2	21	9	66			.1g
Corylus americana		3					
Quercus spp.			3	2			

[a] K. Brown 1981; [b] Blakeslee and Rohn 1982; [c] M. Brown 1982b; [d] Schmits, ed. 1982; [e] Leaf and Root 1979; [f] Root 1981b.

* weights of nutshell fragments are provided when available; otherwise counts are listed.

the various Plains Woodland and Plains Village complexes within the Central Plains. This taxonomic distinction is used in this study whenever spatial and/or temporal dimensions warrant.

The Plains Woodland taxonomic complexes of interest to this study include Grasshopper Falls, A.D. 500-1000 (Reynolds 1979); Greenwood, A.D. 500-1000 (Witty 1982); Sterns Creek, A.D. 1100-1250 (Haas 1983); Hertha, A.D. 365-760 (Blakeslee and Rohn 1982); Butler, A.D. 500-800 (Grosser 1973; Adair 1983); and Keith, A.D. 600-900 (A. Johnson, in press; Kivett 1949, 1952).

Spatially separated within different geographical areas of Nebraska and Kansas, these cultural phases are best identified by differences in ceramics, particularly surface treatment and temper. Other Woodland complexes have been identified within the Central Plains (Johnson in press); however, adequate information on settlement-subsistence is lacking, making it difficult to include them in the present discussion.

The Hertha phase was defined by Blakeslee and Rohn (1982) for components in the Hillsdale Lake area in the northern part of the Osage Cuestas (Fig. 2.3). Radiocarbon dates place this phase between A.D. 365-A.D. 760 and it is believed to have developed from preceding Early Ceramic complexes. Excavated sites have disclosed small, oval (5 m by 7 m) structures with central rock-lined hearths. Several of these structures have been interpreted as winter habitations (Blakeslee and Rohn 1982). Characteristic artifacts include small to medium corner notched points, scrapers, very few ground stone tools, and cord roughened globular shaped vessels (Brown and Brown 1986).

Subsistence items included deer, bison, rabbit, other small game, and freshwater mussel. Wild plant foods are represented by pokeweed (*Phytolacca americana*), pigweed (*Amaranthus sp.*), vetch (*Vicia sp.*), hickory (*Carya sp.*), kingnut (*Carya laciniosa*), and black walnut (*Juglans nigra*) (Brown and Brown 1986:135).

The Grasshopper Falls phase consists of recognized sites and components dispersed along natural floodplain elevations of the Delaware River and its tributaries in northeastern Kansas (Fig. 2.3) (Reynolds 1979; Williams 1986). Excavated sites indicate a settlement pattern composed of semi-sedentary habitations of wattle and daub construction. Individual sites contain as few as one or two houses and occasionally as many as 12 structures. House size encompasses from 35 to 136 sq m and probably reflects occupation by a nuclear family (Reynolds 1979). Features such as storage pits are found both within and without the houses while hearths are generally extra-mural.

Recovered artifacts include styles associated with hunting and butchering and grinding stones suggest the addition of plants in the diet. If present at all, horticulture was of minimal importance (Reynolds 1979:73). Excavations at the Woodland component at 14AT2, yielded identifiable remains of goosefoot (*Chenopodium sp.*), pigweed (*Amaranthus sp.*), purslane (*Portulaca sp.*), bulrush (*Scirpus sp.*), and black walnut (*Juglans nigra*) (Williams 1986:46).

Geographically located within the Flint Hills and the western Osage Plains of eastcentral Kansas is the Greenwood variant (Fig. 2.3). Data from excavations indicate a similar settlement pattern as that presented for the contemporaneous Grasshopper Falls phase. Houses are identified as round to oval structures having interior and exterior features (Reynolds 1984). Research has identified the presence of both large villages with numerous house remains as well as smaller shorter-lived camps (Witty 1982), suggestive of both social aggregation and segmentation practices. Subsistence remains include a variety of large to small game animals such as bison (*Bison bison*), deer (*Odocoileus sp.*), elk (*Cervus canadensis*), raccoon (*Procoyon lotor*), beaver (*Castor canadensis*), coyote (*Canis lupus*), and rabbit (*Sylviligus floridanus*). Floral remains are poorly represented, partially due to the lack of systematic recovery techniques for these particles during excavation. Recent investigations at the Stiles site (14MM13) however, recovered botanical remains from floating feature fill and identified plants include hickory (*Carya spp.*), walnut (*Juglans nigra*), hazel (*Corylus americana*), vetch (*Vicia sp.*), goosefoot (*Chenopodium sp.*) and one corn kernel fragment (*Zea mays*) (Blakeslee and Rohn 1982) (Table 2.5).

The Butler phase (Fig. 2.3), originally defined from excavations at the Snyder site (14BU9) in southcentral Kansas (Grosser 1970, 1973), is now more clearly recognized. The presence of zoned dentate stamped ceramics on the surface of Snyder was inferred as Middle Woodland Hopewell influence and led to the suggested temporal span of A.D. 200-800. Additional subsurface ceramics however were more indicative of a later Plains Woodland occupation. Recent excavations at Snyder and an extensive reexamination of the Woodland materials from the earlier 1968-1972 excavations, along with investigations at other Woodland sites in the area, have identified the Butler phase as a major Plains Woodland occupation in the lower Walnut River valley. A precise location of the Hopewellian materials at Snyder could not be documented either by recent investigations or by an examination of earlier field notes (Gary Leaf, personal communication). The fact that these materials were donated to the Museum by a local amateur and that similar materials have not been found in the valley area warrants a redefinition of the Butler phase (Adair 1983). Additionally, thick limestone or clay tempered pottery and large and small corner-notched dart and arrow points recovered from sites 14BU4, 14BU30, 14BU31, 14BU32, 14BU19, and 14BU57 are similar to materials from Snyder. Radiocarbon dates of A.D. 735±55 and A.D. 710±75 from 14BU31 (Leaf 1979) suggest a tighter temporal span than originally suggested for the Butler phase; perhaps A.D. 500-800 would be more appropriate (Adair 1983).

Data pertaining to the settlement-subsistence practices of the Butler phase are largely the result of recent and ongoing research. A hunting and gathering subsistence mode is accepted to be the common strategy although a single corn kernel has been recovered from 14BU57 (M. Brown 1982b and Table 2.5), indicating at least the introduction of this cultigen into the area. Based on data from Snyder site, post mold configurations exhibit the presence of two, and possibly more, houses measuring approximately five to six meters in diameter (Gary Leaf, personal communication). The houses are outlined by paired posts and often overlap adjacent structures, indicative of repeated occupations. Storage pits are present both inside and outside the structures but hearths are so far absent. Data would therefore indicate a semi-sedentary occupation by groups of nuclear families who followed a hunting-gathering economic mode.

Extending westward from northcentral Kansas and southcentral Nebraska to the western portions of these states and distributed in the north from the Platte River to the Arkansas in the south, is the Keith variant (Fig. 2.3). Known sites of this complex include small temporary camps with extremely limited quantities of cultural debris, villages containing several domestic structures, burial mounds, and ossuaries (Kivett 1949, 1952; Witty 1966; Wedel 1959; Grange 1980; A. Johnson, in press). The sites are primarily located along major rivers and smaller tributary streams of the western Plains. Most of the Woodland components assigned to the Keith variant were excavated in the 1940s-1950s.

The settlement pattern of this variant is poorly understood. Johnson (in press) has suggested that the variability in site types may be indicative of spatially and/or temporally distinct adaptive patterns. However, acceptable radiocarbon dates are very few. Excavations at the Doyle site (25RW28) in the Red Willow reservoir have provided dates of A.D. 520 and A.D. 580, the Woodruff ossuary in northcentral Kansas has been dated to A.D. 607, while dates of A.D. 820 and A.D. 690 are available from two village sites in southcentral Nebraska (Table 2.2). Recent investigations in the Harlan County Lake vicinity have provided two additional Keith variant dates from site 25HN12. These dates are A.D. 700 ± 110 and A.D. 900 ± 70 (Adair and Brown 1986). Given the ceramic and projectile point styles, it seems likely that this Plains Woodland variant dates from about A.D. 600-900 and may represent the earliest semi-sedentary occupation in the western Plains. The variability in site types has not yet been associated with any temporal division.

At present, the best description of domestic structures is provided by Grange (1980) from excavations at the Doyle site. Round basin-like pit houses about 4 - 5.5m in diameter were identified during salvage excavations at the Red Willow reservoir in the early 1960s. These structures contained interior hearths, concentrations of shell and burned rock, extensive amounts of ceramics, chipped stone tools, faunal remains, and subfloor burials. The houses lacked evidence of posts or superstructures which, in conjunction with superimposed hearth remains, led the excavator to conclude that the remains represented repeated occupations. A semi-sedentary settlement pattern is consistent with other Plains Woodland variants to the east. Additionally, a single bison scapula tool exhibiting extensive polish suggests that plants, and possibly cultivation, were also important to these more western Plains populations. If the Doyle site can be considered representative of this complex, the Keith variant differs in regards to faunal exploitation. The variety and number of animal bones recovered is far greater than any other known Plains Woodland site. Identified were remains of bison, deer, antelope, cottontail, jackrabbit, badger, beaver, raccoon, coyote, turtle, Canada goose, eagle, hawk, raven, owl, turkey, teal, mallard, and freshwater mussel (Grange 1980).

Most research on the Keith variant has focused on the burial practices (Kivett 1953; Wedel and Kivett 1956) and on the association with the more northern Valley variant (A. Johnson, in press). Stylistic variability in the ceramics and projectile points already suggest the presence of more than one cultural phase for this complex. Late Keith variant sites, identified by the presence of small, triangular, corner-notched and unnotched points with concave bases, may be important transitional markers from Plains Woodland to Plains Village traditions. Hopefully, future investigations on the Keith variant will elucidate the subsistence strategies, settlement pattern, and temporal range.

A single Woodland component deserves mention in this overview. It is the Walker-Gilmore site (25CC28), located in eastern Nebraska at the junction of the Platte River with the Missouri. Excavations at this site in the 1930s recovered materials used to identify the Sterns Creek variant, a complex believed to spatially occupy an area along the Missouri and its tributaries in eastern Nebraska and western Iowa (Fig. 2.3). The most recent discussion of the type site is based on results of excavations conducted in 1968 (Haas 1983). This report discusses the sedentary lifestyle expressed in four of the five Woodland zones. Remnants of structures, storage facilities, and feature patterns, along with ceramics and chipped stone tools combine to suggest a major and repeated occupation at the site for several centuries. Faunal remains indicated the exploitation of mammals, aquatic resources, and migratory waterfowl while the floral remains included charred squash and gourd seeds, providing some of the best empirical evidence for cultivation during the Woodland period. The recent investigations have suggested that the occupation is somewhat later than originally believed. Acceptable radiocarbon dates (Table 2.2) range from A.D. 1140 to A.D. 1285 (Haas 1983). A previous date from the site, A.D. 920 ± 150, is not in agreement with the new assays which suggest an occupation from the latter part of the eleventh century to the beginning of the thirteenth century.

The trend to increased sedentism during the Plains Woodland period has long been recognized (Wedel 1961, 1986; Kivett 1970). In addition, this period appears to be a time of increased population and increased use of plants in the diet. The settlement patterns of the Grasshopper Falls, Hertha, Greenwood, Butler, and Keith variants indicate a fairly sedentary lifestyle with at least part-time residence each year in small nuclear family size houses. Small campsites attributed to the Keith variant may further suggest that seasonal or yearly hunting forays were characteristic of populations in the western portion of the study area. While population estimates are unavailable for this time period, a general understanding of this factor can be gained by examining site survey records. Table 2.6 lists the number of Woodland components identified from several surveys conducted in eastern Kansas where the cultural affiliation of these sites was determined by the presence of diagnostic artifacts on the surface. Recognizing the possibility of buried sites, Woodland sites tend to comprise the largest class of culturally identified occupations. In some surveys, however, such as the Stranger Creek drainage, the majority of the sites could not be placed in a temporal framework due to the lack of diagnostic materials. Consequently, the percent of Woodland period sites appears low when compared to other surveys. While the individual houses and site components of identified Woodland occupations are relatively small when compared to the Kansas City Hopewell villages, the number of sites is actually quite abundant, suggestive of a larger and more dispersed population than known from preceding periods. Additionally, they are frequently represented by remains of semi-sedentary structures. Some researchers argue that increased sedentism, such as that observed during the Plains Woodland period, would indirectly cause an increase in population (Binford and Chasko 1976; Cohen 1977; Lee 1972; Frisch 1978) and that this population growth may stimulate subsistence change (Asch et al. 1972; Boserup 1965; Cohen 1977; Styles 1981).

Table 2.5 summarizes the available data on probable plant foods utilized by populations inhabiting eastern Kansas and the Kansas City locality during the approximate period of A.D. 500-900.

Bearing in mind that Plains Woodland sites such as 14MM13 and 14BU57 may have developed out of earlier Archaic complexes, the table demonstrates an increased use of plants in the diet from the preceding Archaic period.

Middle Ceramic

The change in subsistence from the Woodland period to following complexes is believed to have involved the adoption of cultivated plants. While this is the subject of Chapter 4 in this study, two cultural complexes are of importance for this chapter in that they temporally span the period of this adoption. The first is the Bemis Creek phase (Fig. 2.4) of southcentral Kansas, identified from excavations at the Two Deer site (14BU55) (Adair and Brown 1981; Adair 1982). Recovered artifacts, including ceramics and chipped stone tools, are reflective of styles associated with both Plains Woodland and Plains Village complexes. In conjunction with radiocarbon dates averaging at A.D. 1000±25 (Table 2.2), these stylistic artifacts led the excavator to interpret the occupation at Two Deer as transitional Plains Woodland/Plains Village.

The sedentary occupation of this site was preserved within the remains of two large structures measuring approximately 10 m in diameter. Within the parameters of one wattle and daub structure were a centrally located hearth and two grinding slabs. Faunal elements scattered throughout the house and around an exterior hearth included remains of deer, bison, beaver, and raccoon. Large scale water flotation of soil samples from midden fill and features yielded an impressive amount of charred seeds and nutshell fragments. Primary flora exploited included goosefoot (*Chenopodium spp.*), pigweed (*Amaranthus sp.*), spurge (*Euphorbia spp.*), grape (*Vitis sp.*) and black walnut (*Juglans nigra*) and various species of grasses and sedges. Cultivated corn (*Zea mays*), squash (*Cucurbita pepo*), sunflower (*Helianthus annuus*) and marshelder (*Iva annua*) were also substantially present in the assemblage (Table 2.7). The co-occurrence of native and tropical cultigens indicates a well developed horticultural subsistence base, yet there is no evidence to suggest that Two Deer represents an occupation of migrating people who brought agricultural foods with them to the Lower Walnut valley.

Table 2.6. Number and percent of Woodland compponents identified by surveys conducted in the study area.

Survey Area	*Number of Woodland Components*	*%*	*Percent Unknown*
Delaware River Valley[a] northeast Kansas	124	61	14
Wanut River Valley[b] southcentral Kansas	27	41	36
Stranger Creek[c] northeast Kansas	51	29	46
Clinton Lake[d] northeast Kansas	28	32	47
Hillsdale Lake[e] eastcentral Kansas	27	23	?

[a] Reynolds 1979; [b] Root 1979; [c] Logan 1983; [d] Chambers and Thompkins 1977; [e] Rohn and Woodman 1976.

The second complex is the Pomona variant of the Osage Plains in eastern Kansas (Fig. 2.4), identified by sites representing a late survival of Plains Woodland manifestations with some shared traits from contemporary Central Plains complexes (Witty 1978).

A detailed synthesis of 34 excavated Pomona occupations in Kansas and seven in Missouri defined four regionally and temporally limited phases within the Pomona variant (Brown 1984). These phases include Clinton, spatially restricted to the Verdigris, Neosho, Marais des Cygnes. Wakarusa, and Little Blue (Missouri) Valleys in the study area and temporally spanning A.D. 960—A.D. 1430. The Wolf Creek phase is identified from sites in the Neosho and Verdigris River drainages which date from A.D. 980—A.D. 1325. Chronologically placed from A.D. 1150—A.D. 1285 and recognized in the Little Blue drainage and Truman Lake area of Missouri and infrequently in the Neosho and Verdigris River drainages of Kansas is the May Brook phase. The fourth phase, Apple Valley, is recognized in the Delaware, Verdigris and Neosho River drainages and dates from A.D. 1300—A.D. 1350. Surface decoration and temper in the ceramics and raw material type of the lithics were the major attributes used to denote phase content (Brown 1984). For purposes of this study, the May Brook phase has been spatially limited in Figure 2.4 while the other three phases have been grouped together under the designation Pomona. Diagnostic artifacts of the variant include small corner-, side-, or un-notched arrow points and globular shaped ceramic vessels with cord roughened or smoothed exterior surfaces. Occasionally, decorations along with strap handles and collared rims are found on the ceramics which is interpreted as influence from adjacent Central Plains populations. Pomona sites are usually identified by the remains of several structures that differ from other Central Plains complexes in their flimsy design, lack of the four central support posts, and lack of an interior hearth (Witty 1978). Semi-sedentary occupation of these structures is inferred by the amount of cultural debris and the presence of storage pits.

Unfortunately little flotation has been conducted at Pomona sites, significantly hampering a discussion of subsistence strategies. Preserved faunal remains suggest a hunting pattern similar to the ancestral Woodland period and limited cultivation has been inferred by the presence of horticultural implements (i.e. scapula hoes) at several Pomona sites. Additionally, cultivated corn (*Zea mays*), squash (*Cucurbita pepo*) and sunflower (*Helianthus annuus*) have been identified from Hatcher (14DO19), located in the Wakarusa River valley. A single corn kernel was identified from the Pomona occupation at site 14AT2 in the Delaware River drainage (Williams 1986); domesticated *Iva* was recovered from the May Brook component at Vaughn Estess (Voigt 1985 and Table 2.7), and corn, beans, and sunflower are known cultigens (Table 2.7) recovered from two late Pomona occupations in the Wolf Creek area (Rohn et al. 1977) (Table 2.2).

Within Kansas and surrounding areas in Nebraska, Missouri, and Iowa, cultural manifestations dated from A.D. 1000-1500 are assigned to the Central Plains or Village Farming tradition of the Plains Village period (Lehmer 1971; Chapman 1980). Included are the Upper Republican, Smoky Hill, and Nebraska complexes (Fig. 2.4). Highly characteristic of these three complexes are the remains of earthlodges, usually identified with four central support posts and a large interior hearth (Wood 1952). Complexes affiliated with these three in northwestern Missouri, southwestern Iowa, and southcentral Kansas are the Steed-Kisker, Glenwood, and Bluff Creek, respectively (Fig. 2.4). Initial similarities between sites of all these complexes are found in the subsistence strategies, with cultigens being ubiquitous in features and on house floors. Differences between the complexes are best noted by variations in ceramics, particularly in decoration and design along with size and architectural styles of residences.

Although a large number of Central Plains tradition sites have been excavated, comparatively few have been described in the literature. This is due in large part to the fact that many of the

Table 2.7. Probable Plant Foods from Plains Village, Pomona, May Brook, Smoky Hill, Steed-Kisker, and Glenwood sites.

	Two Deer 14BU55[a]	*Hatcher* 14DO19[b]	14BU71[b]	*Roth* 14BO319[c]	*Winn* 14CF506[d]	*Anderson* 14CF508[d]	*Seven Acres* 23JA115[e]	*Maybrook* 23JA43[f]	*Bowlin Bridge* 23JA38[g]	*Vaughn Estess* 23JA269[k]	*Young* 23PL4[h]	23CL115[h]	*Hulse* 23CL109[h]	*Coons* 23PL16[h]	*Friend and Foe* 23CL113[i]	23CL226[j]	23CL274[j]	23CL276[j]	*Witt* 14GE600[m]	13ML129[l]	13ML138[l]
Chenopodium spp.	2580			8			22	173	4	3		70	15		x		13	1	11	80	
Amaranthus sp.	876						8	54	21						x						
Polygonum sp.	4			8				1				1	10								
Iva annua	26*									3*	4*	19*		3*	10*		14*	1*	12*	392*	12*
Vitis sp.	11			2		x		1			1									3	
Viola sp.	16							1	13												
Galium sp.	20			4			81			2											
Helianthus annuus	195*	8*	1*			5*					57*	2*	1*	1*					5*	466*	23*
Euphorbia spp.	241						17		1												
Rubus sp.							3	1													
Portulaca sp.	31						30	92	8												
Ambrosia spp.	7						4														
gramineae	139						9	1		1											
Crataegus sp.											2		10								
Cucurbita pepo	31	3												2	2					35	114
Lagenaria siceraria												1		3							
*Zea mays***	114	2			x	x	2		1		144	33	121	249	38+	3	3	4	110	105	224
Phaseolus vulgaris					x	x						19	17	134						18	65
*Juglans nigra** ***	75.9 g					x	478	5	2	20	x	5	25		x		2.0	.2	6	12.0	2.0
Carya spp.	4.4 g			2		x	168	30	11	284	x	14	42	4	13+	0.1	6.8	1.2	11	14.0	2.5
Corylus americana							1			1	x	3	30	3	2	0.1		.1			
Quercus spp.	8.5 g						136	168									1				

[a] Adair 1982; [b] Adair, unpublished notes; [c] Brogan 1982; [d] Rohn et al. 1977; [e] K. Brown 1981; [f] Schmits 1982; [g] Schmits (ed.) 1982; [h] Blake 1974, 1980; [i] Calabrese 1969; [j] McHugh 1982; [k] Voigt 1985; [l] Nickel, n.d. and Adair, unpublished notes; [m] Adair 1984b

* represents a cultigen variety

** includes grains, cupules, and cob fragments

*** weights of nutshell fragments are provided when available, otherwise counts are listed

x present but not quantified in manuscript

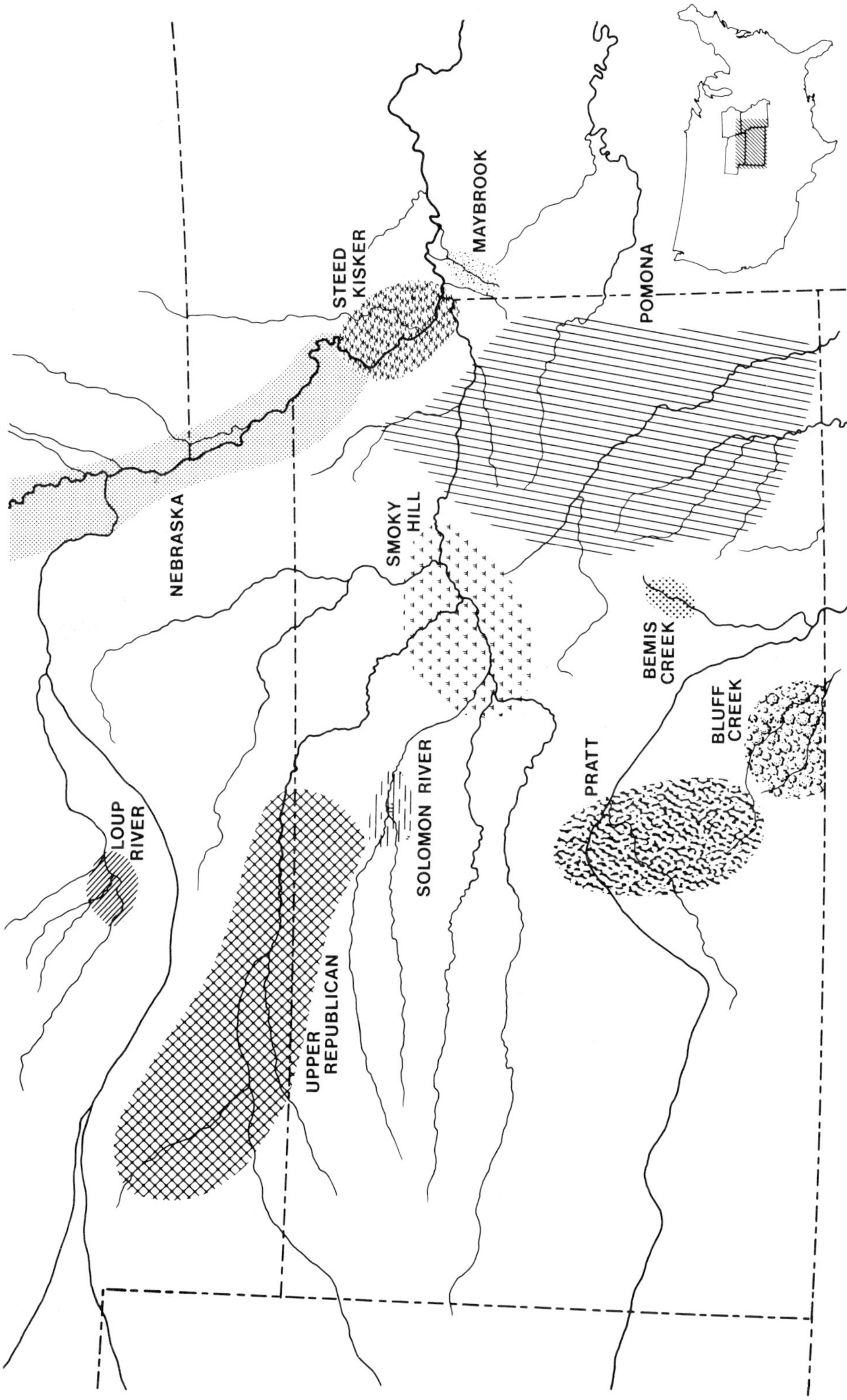

Fig. 2.4 Location of Middle Ceramic complexes within the study area.

sites were investigated in the 1930s with the use of WPA crews and in the 1950s in conjunction with River Basin Surveys and the National Park Service. Salvage work during these times was not always followed by a descriptive report. Much of the data collected during these times however have remained in institutions within the Central Plains which allowed for subsequent analyses and reports (e.g. Blakeslee and Caldwell 1979; Johnson 1973; Lippincott 1978). Reports such as these were consulted for the following discussion.

The Upper Republican complex is geographically positioned in northwestern Kansas and southwestern Nebraska, primarily the areas occupied by the earlier Keith populations. While there is no cultural connection known from Keith to the Upper Republican variant, the latter has been taxonomically divided into three developmental phases, the first of which shows similarities to the earlier Plains Woodland variant. Initial occupation in north-central Kansas along the Solomon River drainage by Central Plains people has been termed the Solomon River phase (Lippincott 1978). It is identified by remains of individual houses, small farming communities consisting of several structures, seasonal campsites, and burial ossuaries; a site variability similar to the Keith variant. Radiocarbon dates indicate a temporal span of A.D. 850-1250 (Table 2.2). The Classic Republican phase overlaps Solomon River by 200 years and continues along the Republican River and its tributaries in western Nebraska and Kansas until approximately A.D. 1350 (Ludwickson 1978). The final phase, Loup River, dated from A.D. 1250-1500, represents an "amalgamation of predominantly Upper Republican phase people with lesser elements of Smoky Hill and/or Nebraska phase people...and is a linear antecedent of the protohistoric Pawnee Lower Loup phase" (Ludwickson 1978: 94). It is spatially restricted to areas surrounding the lower reaches of the Middle and South tributaries of the Loup River in central Nebraska (Fig. 2.4).

Several major studies of the Upper Republican complex have centered on the settlement pattern in the Glen Elder reservoir area, or what basically constitutes the Solomon River phase. Krause (1969, 1970) suggested that hamlets, consisting of several small houses, occurred early in the sequence while larger isolated households and seasonal hunting camps represented later occupations. Subsequent studies in this area (Lippincott 1978), involving a detailed ceramic seriation, suggest instead that the different site types span the entire period of occupation in the area and conclude that "large houses are early and small houses are early and about the only thing that is late is a lack of houses" (Lippincott 1978:90). The author recognizes that different house sizes may reflect different social gatherings but he offers no explanation for the Solomon River sites.

Like the preceding Keith variant, faunal remains at Upper Republican sites are indicative of an exploitation pattern focused on heavy utilization of the local environment. Falk (1969b) has identified 32 taxa of animals from the Mowry Bluff site (25FT35) in southwestern Nebraska. Included are remains of bison, deer, antelope, cottontail, jackrabbit, beaver, fox, prairie dog, Canada goose, teal, turkey, hawk, and several species of reptiles and fish. The faunal remains are indicative of intense utilization, a practice that Falk (1969b) suggests could not be maintained for an entire year by the existing Upper Republican populations. The presence of small hunting camps in the Nebraska

Panhandle, in southeastern Wyoming, and in northeastern Colorado prompted the suggestion that these people engaged in major communal hunting forays, a strategy that would alleviate any pressure caused by over-exploitation of the riverine valleys surrounding the habitations.

Agriculture was a significant subsistence strategy during every phase of the Upper Republican. Charred remains of corn, squash, beans, sunflower, and marshelder (Cutler and Blake 1969, 1973b; Witty, personal communication; Chapter 3, this study) have been recovered from intra-mural features. Additionally, the presence of agricultural implements, such as bison scapula hoes, implies the importance of agriculture from sites where floral remains have not been recovered. Sites discussed in this study that have associated cultigens are the Solomon River phase components at 14ML371, 14ML15, 14ML16, 14ML417, 140B27, and 14JW301; and three Classic Republican components at 14CY102, 14NT301, and 25FT35.

The Nebraska variant was first defined in the 1920s by an amateur archaeologist conducting excavations around Omaha (Gilder 1926). Investigations continued for many following decades by the University of Nebraska which subsequently led to the introduction of classificatory schemes for the Nebraska culture. In 1969, Krause proposed two developmental phases, an earlier Doniphan phase positioned along the Missouri River in northeastern Kansas and southeastern Nebraska, and a later Douglas phase located north of the Platte River in eastern Nebraska. An extensive examination of published and unpublished documents on Nebraska phase sites led to the most recent definition of this cultural complex (Blakeslee and Caldwell 1979). An analysis of ceramics, architectural style, house size, and radiocarbon dates led the authors to suggest the presence of two distinct populations in eastern Nebraska: "a population which moved north along the Missouri River bluffs from Kansas to northeastern Nebraska and a population which moved into the tributary stream valleys to the west several hundred years after the first population had staked its claim to the Missouri River bluffs" (Blakeslee and Caldwell 1979:118). Still another population may be responsible for the remains in the Glenwood locality of western Iowa. Temporal dimensions for sites in Nebraska range from A.D. 780 ± 60 (25RH1) to A.D. 1660 (25DO2), although dates earlier than A.D. 1000 and later than A.D. 1435 have been dismissed since they are "obviously in error" (Blakeslee and Caldwell 1979:19). Nine localities within the Nebraska variant were also recognized. Included are (from south to north) Brownville, Nemaha, Plattsmouth, Weeping Water, Salt Creek, South Bend, DeSoto, and Macy in Nebraska and Glenwood in Iowa. The Nemaha, Salt Creek, and South Bend are located on western tributaries of the Missouri while the remaining localities are positioned along this river.

An examination of architectural styles (i.e. house shape and size, position of central support posts, length of entranceway), radiocarbon dates and a seriation of ceramics from sites in Nebraska delineated several characteristics of this complex (Blakeslee and Caldwell 1979). There appears to have been a south to north movement of Nebraska culture with later period sites containing a greater percentage of decorated ceramics and often displaying circular rather than square domestic structures. Both single households and villages occur throughout the spatial and temporal lifetime of the variant although, as in the case of the Solomon River phase, single lodges tend to be considerably larger than those in a village setting. Shell tempered ceramics occur early in the sequence and are recovered mainly from sites in northeast Kansas and southeast Nebraska. Nebraska phase occupations on western streams tributary to the Missouri River apparently begin around A.D. 1200 and are viewed as being "ceramically distinct" from the earlier populations (Blakeslee and Caldwell 1979:116). A further suggestion that this population, which produced ceramics with collared rims and no decoration, was connected with the more southern Smoky Hill phase, awaits clarification.

While this most recent appraisal of the Nebraska variant is extremely thorough in its treatment of ceramics and settlement patterns, it is of interest to note that nowhere do the authors discuss the subsistence mode. It is generally assumed that hunting, gathering, and agricultural strategies were followed although very limited information is currently available on the kinds and amounts of plants collected or grown. Considering when most sites were excavated, this problem is somewhat expected and accepted. For example, Heavin (1970), in her report on two large storage pits at the Squaw Creek site (14DP13), includes personal communication from J. Mett Shippee, who remarked that he recovered about one gallon of charred shelled corn from his 1959 excavations at the site. These remains, along with the recovered common beans, are the only ones from Nebraska phase sites south of the Platte River included in Cutler and Blake's (1973b) extensive review of cultigens in the area. Interestingly, Wood (1969b) proposed the likelihood that agriculture was less environmentally risky for Nebraska populations than for Upper Republican complexes, a factor that may have prompted more intensive and hence more productive horticulture in Nebraska communities. This, combined with the area's rich and varied natural resources and a lack of known hunting camps, suggests that Nebraska populations had a far more sedentary way of life than the Upper Republican peoples (Wood 1969b:104). The most likely explanations for the lack of floral remains that would substantiate this suggestion are that they were either not collected during excavation or that they have never been analyzed and reported. Given the extensive botanical work done on the Glenwood locality sites (discussed below), it seems probable that Nebraska populations cultivated and collected similar plants.

The Glenwood locality is identified in westcentral Iowa by remains of single lodges and villages and shares several characteristics with the Nebraska variant. While Blakeslee and Caldwell (1979:22) actually refer to the culture as the Nebraska phase in Iowa, Hotopp (1978) retains it is a separate complex with the taxonomic distinction of "locality." Only further comparisons between the two complexes can resolve this issue.

Glenwood sites were first identified in the late 1880s and have continued to be the focus of research since that time. They are commonly located along the Missouri River bluffs, in proximity to southwesterly flowing drainages as Pony, Keg, and Horse Creeks (Fig. 2.4). Variability in sites is noted by the presence of single large lodges, villages of 2-5 lodges, and houses located in a linear pattern along ridges but not organized into a village or community pattern. Like Solomon River and Nebraska, these sites occur contemporaneously as the radiocarbon evidence does not support a shift from hamlets to homesteads. Additionally, radiocarbon dates from sites recently investigated in the Pony and

Keg Creek drainages suggest that the Glenwood settlement did not proceed sequentially through an extended range as originally proposed by Anderson (1961) but rather existed as intensive occupations in both watersheds for several centuries (Hotopp 1978). Approximately 76% of the radiocarbon assays are contained within a 300 year span from A.D. 1000-1300 (Tiffany 1981; Table 2.2, this study).

Like other Central Plains tradition cultures, Glenwood subsistence was based on mixed strategies of hunting, gathering, and agriculture. However, unlike other Central Plains tradition cultures, the latter strategy has been clarified with varieties of floral remains quantified (Nickel n.d. Adair, this study). Nine of the eleven sites investigated in Pony and Keg Creeks during a recent highway relocation project contained cultigens, including corn (*Zea mays*), squash (*Cucurbita pepo*), beans (*Phaseolus vulgaris*), sunflower (*Helianthus annuus*) and marshelder (*Iva annua*) (Table 2.7). Additional floral remains gathered from local resource zones include hickory (*Carya sp.*), black walnut (*Juglans nigra*), hazel (*Corylus sp.*), cherry and plum (*Prunus* cf.*virginiana* and *americana*), grape (*Vitis sp.*), goosefoot (*Chenopodium sp.*), and hackberry (*Celtis sp.*).

Components assigned to the Smoky Hill variant that are included in this study are Budenbender (14PO4), Walls (14MN42), Griffing (14RY401), Witt (14GE600), Root (14LC301), Minneapolis (14OT5), Mugler (14CY1), Kemp (14CY4), Woods (14CY30), 14GE21, 14GE127, and one unnumbered site from the Schultz collection. These sites, and others assigned to the Smoky Hill variant, are clustered around the lower Kansas, Republican, Solomon, and Smoky Hill Rivers (Fig. 2.4). Lodges are usually located on well drained terraces or bluff tops overlooking alluvial valleys. Villages are described as small hamlets with one or two house units, widely scattered structures strung along a creek terrace or bluff top, or a large farmstead community consisting of perhaps as many as 15 house structures (Wedel 1959: 513, 563). Radiocarbon dates (Table 2.2) indicate a temporal range of A.D. 1000-1350 and it is assumed that the variability in site types spans the entire temporal range, as demonstrated for the Upper Republican and Nebraska variants.

Components of the Smoky Hill variant were first labeled as taxonomically distinct from other Central Plains complexes by Wedel's work in the area around Salina, Kansas (Wedel 1935, 1959). Excavations at the Whiteford and Griffing (14RY401) sites were instrumental in defining this cultural complex. Subsequent excavations at Smoky Hill sites (A. Johnson 1973; Witty 1962, 1963; O'Brien, personal communication) have aided in recognizing diagnostic traits and variability within this variant.

Wedel (1959:563) describes the lodges as square to rectangular structures, exhibiting four central support posts, an interior hearth, and an extended entryway. From excavations at Griffing and Minneapolis sites, he also concluded that the structures were built on ground level rather than subterranean, as the Upper Republican and Nebraska earthlodges. These may not be exclusive traits, however, since Witty (1963) describes a semi-subterranean rectangular house from the Woods site and Johnson (1973) presents evidence of a semi-subterranean circular structure at Budenbender. While architectural variability apparently exists, the Smoky Hill variant retains characteristics that separate it from contemporaneous Central Plains complexes, the most noticeable of which are Middle Mississippian traits.

Included are projectile points with multiple pairs of side notches, clay pipes and effigies, and shell tempered pottery (Brown 1967; Wedel 1959; A. Johnson 1973). Wedel (1959:565) also calls attention to the possible contact with southeast Caddoan populations, noting exotic "Spiro Focus" pottery in association with Smoky Hill burials. Local Smoky Hill ceramics are also distinctive in the presence of clay and bone as tempering agents but share with both Upper Republican and Nebraska such traits as collared rims, decorated rims, finger decorations, and the presence of handles and lugs (Brown 1967).

Subsistence strategies were undoubtedly focused on hunting, gathering, and agriculture although floral and faunal remains are rarely discussed in reports. Witty (1963) remarks on the noticeable absence of bone and shell detritus from the Woods site and suggests that it may be due to soil conditions not conducive to preservation. However, in the most ambitious study of Smoky Hill fauna exploitation to date, Brown (1982a) identified 59 taxa of fish, amphibians, reptiles, birds, and mammals from three Smoky Hill sites. Identified fauna from Witt (14GE600), Griffing (14RY401) and Budenbender (14PO1) included large mammals such as bison, deer, wapiti, and pronghorn; smaller mammals like raccoon, beaver, badger, and cottontail; rodents such as gopher, mouse and rat; amphibians, including toad, frog, and turtle; birds represented by grebe, teal, mallard, goose, and eagle; and fish such as catfish, crappie, drum, and gar. Limited amounts of charred corn have been recovered from several Smoky Hill sites (Cutler and Blake 1973b), but recent examination of the Schultz collection materials has demonstrated that this cultigen was abundantly present in several lodges (Table A.4). Literally jars of charred corn were saved during excavations conducted by Floyd Schultz in the 1920s and 1930s. These remains have been curated by the Museum of Anthropology since they were donated by the amateur archaeologist in the 1950s and are included in this study. Additionally, botanical remains recently identified from the Witt site (14GE600) are also included (Adair 1984b). Remains consist of goosefoot (*Chenopodium sp.*), grape (*Vitis sp.*), black walnut (*Juglans nigra*), hickory (*Carya sp.*), domesticated marshelder (*Iva annua*), sunflower (*Helianthus annuus*), and corn (*Zea mays*) (Table 2.7).

The Steed-Kisker complex is believed to have initially originated by migration of individuals into the Kansas City locality since the phase "has no relationship with the previous local Late Woodland culture" (O'Brien 1978a:72). With the exclusion of the ceramics however, Steed-Kisker artifacts appear very similar to other Central Plains complexes. Components assigned to this phase that are discussed in this study are located in Platte and Clay counties in northwest Missouri. Included are Steed Kisker (23PL13), Friend and Foe (23CL113), Vandiver Mounds (23PL6), McClaron (23PL54), Gresham (23PL48), Butcher (23CL118), Coons (23PL16), Hulse (23CL109), Young (23PL4), White (23PL80), 23CL274, 23CL276, 23CL115, 23CL226, and Calovich (14WY7).

Analyses of the Steed-Kisker phase have focused on the settlement pattern (Calabrese 1969; O'Brien 1978a, 1978b); the diagnostic traits (Shippee 1970, 1972; O'Brien 1976; McHugh 1982); the similarities to Mississippian occupations in the St. Louis area, evident especially in the ceramics (O'Brien 1970, 1972, 1973, 1974, 1976); and the affiliation with the Central Plains Tradition, most notably the Nebraska variant (Calabrese 1969; O'Brien 1970;

McHugh 1982). O'Brien (1978b) has recently identified four functionally different types of sites; habitation areas with houses, storage pits and trash areas; storage sites with pit facilities; hunting and butchering camp sites; and burial mound/cemetery areas. Subsistence information is available from the first three site types, although some floral remains are known from mounds as well. Hunting, gathering, and gardening were the resource strategies of the Steed-Kisker people, the first reflected in the remains of deer, bison, dog, raccoon, fox, rabbit, and freshwater mussels (Chapman 1980: 158). Flotation has documented the selection of a variety of native flora (Table 2.7) as well as cultivated crops of corn (*Zea mays*), squash (*Cucurbita pepo*), gourd (*Lagenaria siceraria*), beans (*Phaseolus vulgrais*), sunflower (*Helianthus annuus*) and marshelder (*Iva annua*) (Blake 1974, 1980). These latter remains were recovered mainly from storage pits within houses excavated at several habitation sites.

Radiocarbon dates for Steed-Kisker are relatively abundant (Table 2.2) and allow for a tight control of the temporal dimension. Given the "non-midden development and the homogeneity of the ceramic designs along with the radiocarbon dates . . . the Steed Kisker complex is relatively short-lived, probably dating about A.D. 1000-1200" (O'Brien 1978a: 69-70).

Calabrese (1969) has suggested that Steed-Kisker was genetically ancestral to the Doniphan phase of the Nebraska variant, a proposition based on changes in ceramic design, architectural shifts, and radiocarbon dates. O'Brien (1978b) refutes this thesis that the Steed-Kisker people ultimately "joined the Central Plains Tradition" and maintains that the data suggest instead a local complex that remained distinct from its contemporaneous Central Plains neighbors while engaging in trade with these northern and western complexes. O'Brien bases her conclusions on ceramic design elements which show no evidence of a continual south to north cline in conjunction with radiocarbon dates. Some interaction besides trade apparently did occur however between Steed-Kisker and Nebraska populations, according to analyses of skeletal remains from Steed-Kisker (Stewart 1943; Barnes 1977) and Nebraska (Nickels 1971) burial mounds. Biological ties, evinced by cranial measurements, were noted among the two populations.

Sites assigned to the Bluff Creek complex are also temporally placed within the Central Plains or Middle Ceramic period (Table 2.2) and are located in southcentral Kansas, mainly Harper and Barber Counties. Investigations by the Kansas State Historical Society in the late 1950s and early 1960s have identified this complex as possessing small village sites with lodge locations visible as low circular mounds (Witty 1978). Recorded and investigated sites included in this study are Anderson (14HP5), Buresh (14SR303), Nulik (14SR305) and Anthony (14HP1).

Excavations have identified square, oval, and rectangular structures that appear to be substantial permanent residences but are architecturally different from the styles discussed above. Artifacts present within these dwellings include stone and bone tools and ceramics that exhibit characteristics of the Washita phase of Oklahoma (Witty 1969; Munsell 1961). Charred remains of corn (Cutler and Blake 1973b; Table A.4 this study) and various animals were extracted from interior features. Since many Bluff Creek sites were excavated prior to the advent of flotation, additional information on subsistence, particularly agriculture, is lacking. Given the presence of permanent dwellings and bison scapula digging hoes however, it may be inferred that horticulture was of significance to the inhabitants and included more than just the cultivation of one crop. The Bluff Creek complex offers promising information on the role of agriculture at A.D. 1000 with particular interest generated by its association with more southeastern populations. Corn recovered from two of these sites, Anderson and Buresh, was analyzed and included in this study (Table A.4).

The final cultural manifestation outlined in this chapter is the Pratt complex. Located in southcentral Kansas within the Arkansas River Lowlands (Fig. 2.4), the complex is a late Middle Ceramic expression related to both Southern Plains Caddo complexes and Central Plains tradition cultures (Wedel 1959). It is believed to date between A.D. 1400 and A.D. 1500 and may be ancestral to the Great Bend aspect. House patterns consist of slightly flattened sides with rounded and braced corners, central hearths, and interior storage pits. The settlement pattern is thought to be permanent year-round occupation within these structures (Brown and Brown 1986). To date, only three Pratt complex sites have been investigated, making settlement reconstruction difficult.

The subsistence economy is believed to have been a combination of hunting, gathering, and agriculture. A few remains of corn have been recovered (Table 3.6) along with bison scapula hoes, a diagnostic tool of Central Plains agriculture. Monger (1970) also lists the recovery of plum pits and berry seeds from the Larned site (14PA307). Wild animals are represented by primarily bison and deer.

Thus, from the Late Archaic period through the Plains Village Tradition, the Central Plains was continually occupied by various prehistoric groups. All confronted the basic climate, vegetation, flora, and fauna of the area, although different adaptations in this region have been noted. In particular, different plant foods have been recognized as varying in importance or utilization through time. In the following chapter, plants designated as cultigens, that have been recovered from sites discussed in this chapter, are presented. The botanical data, along with the cultural and environmental setting of the study area, combine to aid in the development of a model on changing subsistence.

CHAPTER 3

The Botanical Evidence

In this chapter, available data are presented to document the appearance of cultigens in the Central Plains. Both native and tropical plants are considered with an emphasis placed on the importance of native plants in the adoption of agriculture, as these have received little prior attention. Current knowledge of domesticated plants in the Central Plains subarea includes the tropical cultigens of corn (*Zea mays*), squash (*Cucurbita pepo*), bottle gourd (*Lagenaria siceraria*), and beans (*Phaseolus vulgrais*) and the native cultivars of marshelder (*Iva annua var macrocarpa*) and sunflower (*Helianthus annuus var macrocarpus*). Evidence for the domestication of each of these plants will be discussed separately. Ongoing research in various areas of the eastern Woodlands (Asch and Asch 1985; Fritz 1984; Cowan 1985; Smith 1984; Watson 1985) has demonstrated the cultivation and domestication of several other native annuals. In addition to marshelder and sunflower, the list includes *Chenopodium berlandieri* (goosefoot), *Polygonum erectum* (knotweed), *Phalaris caroliniana* (maygrass), *Hordeum pusillum* (little barley), and *Ambrosia trifida* (giant ragweed). Although seeds of some of these species do occur in archaeological contexts in the Central Plains, for reasons discussed below, current data are insufficient to argue for their cultivation. A brief review of these cultigens is presented here along with a summary of their current status in the study area.

At the present time, nomenclatural status for the domesticated archaeological chenopod indicates it belongs to subsection *Cellulata* of section *Chenopodia* (Asch and Asch 1985). The two wild species of *Cellulata* found in North America are *Chenopodium bushianum* and *C. berlandieri*. Differences between these two species may be slight as Asch and Asch (1985;174) prefer to assign prehistoric remains from westcentral Illinois to the exclusive taxon *C. berlandieri* and relegate *C. bushianum* to a subspecies or varietal level. Cultigen chenopod may have been introduced into the southeastern United States from Mesoamerica in late prehistoric times, as noted by Fritz's (1984) identification of *C. berlandieri* ssp. *nuttalliae* seeds from rockshelter sites in northwest Arkansas. This does not preclude the possibility of a previous and independent domestication of an indigenous species of *Chenopodium* however. Asch and Asch (1985:171-183) have argued for domesticated chenopod in Middle Woodland sites in westcentral Illinois and also note the significant utilization of *C. berlandieri* during the Archaic Titterington cultural phase at 2000 B.C. Perhaps the best direct evidence for the intensive use of this plant as a food comes from Kentucky cave and rockshelter sites where chenopod seeds have been identified in human coprolites dated from 710 B.C.-290 B.C. (Yarnell 1969).

As presented in Chapter 2, *Chenopodium* spp. are present, sometimes in abundant quantities, in Central Plains sites ranging in time from Middle and Late Archaic to Middle Ceramic. When given, most species level identifications are either *C. album*, an abundant variety, or *C. hybridum* var. *gigantospernum*, an unusually large fruited species. Far greater attention needs to be given to the subgenus classification of this plant and its prehistoric economic importance. *C. album* is now recognized as an introduced species that closely resembles the most common indigenous species, *C. berlandieri* (Bare 1979). The prehistoric use and possible cultivation of this species in the Central Plains would revolve around a different set of arguments than those posed for the eastern Woodlands where the species is not widely distributed (Asch and Asch 1985, Cowan 1985). It is worthy to stress however that the abundance of the small seeds at the Middle Archaic Coffey site, the Late Archaic Nebo Hill occupation and the Early Woodland Traff site (Table 2.3) clearly indicates an exploitation and heavy utilization of a native annual. With the exception of the Titterington phase occupation at Napoleon Hollow, this practice differs from that observed for the same cultural periods in westcentral Illinois (Asch and Asch 1985).

While *Polygonum erectum* is distributed throughout eastern North America it is a common seed only in flotation samples from prehistoric sites in the lower Illinois River Valley. In this location it is "one of the quartet of starchy seeds which dominate Middle and Late Woodland seed assemblages" (Asch and Asch 1985: 183). The hypothesis for knotweed cultivation beginning during the Middle Woodland period focuses on the plants modern natural distribution and abundance as well as its prehistoric economic status.

P. erectum is not recognized as one of the 18 species of knotweed indigenous to present day Kansas (Bare 1979) and it has not been identified, to my knowledge, from any archaeobotanical collection within the study area. *Polygonum sp.* however is represented in several samples from Archaic to Middle Ceramic period sites (see Tables 2.3, 2.4, 2.5 and 2.7) although in very low frequencies. Thus it is difficult to even argue for the prehistoric harvest of this seed in the study area; we may very well be dealing with a species common to ground around human habitations which may or may not have been deliberately collected.

Like *Chenopodium*, the cultigen status of *Phalaris caroliniana* (maygrass) has recently been reviewed (Cowan 1978). Primarily a southeastern annual species, its cultivation status rests on its prehistoric distribution far north and east of its present day range, its association with other members of the eastern Agricultural Complex, and its sudden abundance in Early and Middle Woodland deposits suggestive of a major change in the economic pattern (Asch and Asch 1985; Cowan 1978, 1985; Watson 1974). Unlike *Chenopodium* however, maygrass does not seem to have been genetically altered by human use and harvest.

Even though maygrass is considered an adventive weed capable of tolerating a variety of habitats and at least one reference (MacKenzie 1902; cited by Asch and Asch 1985:188) lists its range to include western Missouri, it is not considered to be an indigenous plant of the Central Plains environment. If found in the eastern portion of the study area, its occurrence must certainly be rare, as Asch and Asch's (1985:188) attempts to locate it in this region have been unsuccessful. If present in archaeological deposits, and especially if represented in any quantity, an argument could easily be made for its potential cultivation. At the present time however, *Phalaris caroliana* has not been identified from any macrofloral assemblage from the Central Plains.

A fourth starchy seed belonging to the Woodland period horticultural complex has recently been identified as little barley (*Hordeum pusillum*). Its prehistoric use spread from southern Arizona Hohokam to Caddoan populations of Spiro in Oklahoma to Middle/Late Woodland populations in westcentral Illinois (Asch and Asch 1985). In the latter area it is a likely candidate for cultivation status since it appears in sizeable amounts along with other native cultigens in Middle Woodland sites and it provides an economic edge by maturing during the early summer, long before other starchy or oily seeds ripen. Presumably it would have been grown as a food source although processing little barley is apparently more difficult and time consuming than for other plants (Asch and Asch 1985:193).

The distribution of little barley in the Central Plains is quite extensive. The University of Illinois Agricultural Experiment Station (1976) describes it as a common species to all of Kansas, Oklahoma, Missouri and Nebraska (except the extreme northeast portions of the last two states where it is listed as an occasional species). Schroeder (1979:15) lists it as a common plant throughout Nebraska where it is most frequently found in abused pastures, waste areas and along roadsides. In these areas, the wild plant grows from 4 to 16 inches tall, heads in May and June, and the seeds mature in June and July. The small seeds range in size from about 3-7 mm in length and are not easily distinguished from other grass seeds (Asch and Asch 1985:192). This may explain why it is not listed in any Central Plains macrofloral list although the "catch-all" category of *gramineae* is certainly well represented at some sites. In particular, almost 140 grass seeds are reported from the Two Deer site in southcentral Kansas (Table 2.7), representing approximately 3% of the identified remains. At least some of these seeds closely resemble the barley illustrated by Asch and Asch (1985:191) in both morphology and size. The presence of little barley in Central Plains sites clearly awaits further research.

The last native annual that may have reached a state of cultivation in the eastern Woodlands is giant ragweed (*Ambrosia trifida*). Originally labeled a cultigen by Gilmore (1931) after his examination of collections from prehistoric Ozark bluff dweller sites, ragweed's cultigen status was dismissed by Payne and Jones (1962), only to have the issue reopened after samples from Kentucky rockshelter sites indicated the plants close association with other well known cultigens (Cowan 1985). Like maygrass, giant ragweed seeds have apparently undergone no genetic change due to human manipulation. This fact, along with their relatively low representation in macrofloral assemblages makes it difficult to apply a cultigen status to this plant. This does not imply however that it could not have been deliberately gathered or harvested - only that its economic advantages and seed morphology do not argue for its domestication.

Ambrosia trifida is a common annual herb throughout the Central Plains (Bare 1979, Schroeder 1979). It occupies fields, roadsides, waste places and other disturbed areas, often becoming the dominant plant in these locations. Its prehistoric distribution likewise included the study area, as it is identified in several archaeological collections. It is not an abundant plant however, and rarely has the subgenus classification been designated as *trifida*. Like *Polygonum*, these factors make it difficult to present this plant as much more than an accidental occurrence in archaeological deposits.

The archaeobotanical data presented in the remainder of this chapter were gathered from a variety of sources (see Chapter 1). The modern comparative samples, such as sunflower and marshelder, were compiled from two sources. First, when possible, samples were collected by the author from areas ecologically similar to those from where the archaeological site is located. Care was taken to collect achenes from heads or spikes positioned at different locations on the plant. This is particularly important for sunflower since achene sizes vary from the center to the margins of the head and from a main head to those on secondary branches (Heiser 1978). Achenes of *Iva* apparently do not vary a great deal along a single spike (Asch and Asch 1978). Second, herbarium samples from the Biological Survey, University of Kansas, were added as supplements to the modern collections.

All botanical samples, archaeological and modern, were measured using a micrometer under 10× magnification. Length, width, and thickness measurements taken were the maximum dimensions. For sunflower and marshelder the presence or absence of the pericarp is noted. All measurements listed in the Appendix are the actual measurements and current reconstruction formulas were used to provide the corrected mean sizes. It is likely however, that correction values may change again in the future as new data are brought to light on the domestication process of native species. For this reason, the actual measurements are provided in the Appendix, whereas the text contains reference to the reconstructed sizes according to present formulas. Formulas used for reconstructing the size of sunflower and marshelder remains are provided with the discussion of those particular plants.

Cucurbita

Included under this heading are species of two genera in the family Cucurbitaceae, namely *Cucurbita* (pumpkins and squashes) and *Lagenaria* (gourds). Although the two genera are botanically quite distinct, both were cultivated prehistorically, actually providing the earliest well established evidence for horticulture in North America. There are five species of cultivated *Cucurbita*, all native to the New World and one species of *Lagenaria*, native to Africa. Three of the five cultivated species of *Cucurbita* occurred in North America in pre-Columbian times. Included are *C. pepo* (scallops, yellow crookneck, zucchini, and acorn squash; ''pumpkins''; and yellow-flowered ornamental gourds *C. pepo var. ovifera*), *C. moschata* (Kentucky field pumpkin, butternut squash), and *C. mixta* (cushaw). The two other species, *C. maxima* and *C. ficifolia*, were restricted to south and central America (King 1985).

The most common curcurbit found in archaeological sites north of Mexico is *Cucurbita pepo*, the summer squash and pumpkin. The common New England pumpkin, White Bush Scallop squash, Italian zucchini squash, Summer Crookneck squash, and Acorn squash are all cultivars of the species *C. pepo* (Cutler and Whitaker 1961). Archaeological evidence indicates that *C. pepo* originated from wild species of Central America before spreading northward (Pickersgill and Heiser 1977). The oldest dated remains of this cultigen come from the Oaxaca Valley and date 10,750-9200 B.P. (Whitaker and Cutler 1971). The northward dispersal of this plant into North America evidently began near the end of the middle Holocene period (Kay et al. 1980) with the earliest possible carbonized squash rind fragments dated to 7000 B.P. from the Koster site (Asch and Asch 1982, 1985; Conard et al. 1984). Closer to the Plains, water-logged deposits of squash seeds and rind fragments were recovered from the Phillips Spring site and dated to 4300 B.P. (Chomko and Crawford 1978; Kay et al. 1980).

The green striped cushaw squash and winter pumpkins and cymlings are grouped by the taxa *C. mixta*. Like pepo, this species was present in the Techuacán Valley by 7000-5000 B.P. and was apparently derived from a wild species of southern Mexico (King 1985). *C. mixta* had somewhat the same distribution as *C. pepo* although it did not spread northward as far north as early as pepo squash, perhaps due to its preference for warmer climates. It appears in southern Arizona possibly as early as A.D. 700 (Cutler and Blake 1976) but is best documented several centuries later at Montezuma Castle and Tonto in Arizona (Bohrer 1962; Ford 1981). It has been identified in archaeological contexts from sites in southwest Texas and as far north as southern Nevada and Utah (Ford 1981:19).

Another group of winter squashes and pumpkins is identified as *C. moschata*, although it is better adapted to warm damp tropical environments than the *C. mixta*. Its main areas of distribution include southern Mexico, Central Mexico, Columbia, and Peru. It is also found in the American Southwest by 1250 B.P. (Whitaker and Cutler 1965; Whitaker 1981) although its distribution was limited.

For the Central Plains study area the relevant species of cucurbits is *C. pepo* since it is to date the only one known prehistorically in the area. Therefore it will be the only species of squash under consideration in this study.

The bottle gourd, *Lagenaria siceraria*, was widely distributed over the Americas in pre-Columbian times, and like the squash and pumpkin, it appears to have been gathered and used prior to the adoption of a full agricultural base. In North America, the earliest remains of *Lagenaria* belong to the Early Eastern Mexican Agricultural Complex (Ford 1981) and date to ca. 2500-1500 B.C. Gourd was identified in Archaic deposits at the Iddins site in Tennessee, Salts Cave in Kentucky, at the Phillips Spring site in central Missouri, and at the Cloudsplitter Rockshelter in Kentucky (Cowan 1985; Ford 1981; Yarnell 1969). Additional gourd remains are reported from Late Archaic and Early Woodland contexts in the eastern Woodlands.

Ford (1974, 1981) suggests that squash and gourd husbandry were perhaps disseminated northward from northeastern Mexico into the eastern United States where they were grown in gardens as a complement to a pre-existing hunting and gathering economy focused on climax forest products. It has also been suggested (Whitaker and Bemis 1975) that both pepo squash and gourds existed in eastern North America for a long time as 'weeds' or campfollowers growing on trash heaps and only secondarily dependent upon humans for survival. Likewise controversial is the route followed in the northward diffusion of these plants, an issue hindered by the lack of botanical remains from sites in the southeastern United States. However, the pathway is believed to have been totally independent of the southwestern United States since cucurbit remains in the eastern Woodlands predate those from the Southwest by about 3000 years (King 1985). The importance of the cucurbits as early subsistence items in the eastern Woodlands is not clearly established however. Asch and Asch (1982:3) recognize the earliest cucurbit remains from Koster as a variety of pepo squash but qualify their identification by stating "if one distinguishes between the common usage of *squash* for cucurbits having an edible fleshy rind and *gourd* for cucurbits with an inedible, often hard, and thin shell, then the early material was more gourd-like than squash-like." King (1985:79) indicates that it is almost impossible to distinguish thin *C. pepo* rind fragments from those of *C. foetidissima*, a wild species found throughout the Midwest. The cultivation of a thick fleshier rind squash may have coincided with or followed a selection for larger seed size (Kay et al. 1980; Whitaker and

Bemis 1964). The use of early squash or gourd as a container in preceramic complexes may be an acceptable explanation since additional domesticated plants do not appear in the archaeological record in the eastern Woodlands until 4250 B.P. (gourd at Phillips Spring) and 3920 B.P. (*Iva* from Napoleon Hollow).

Whereas the bottle gourd is considered to have little nutritional value, the other cucurbits are actually an excellent source of food energy. According to ethnographic accounts, fruits of the cucurbits were an important green vegetable, and when mature, they were boiled or roasted as a source of sugary and starchy food (Whitaker and Cutler 1965). Squash seeds alone have a food value of 533 calories/100 g, a value close to hickory nuts (Kay et al. 1980:817). While squash flesh has relatively little food value due to its high water content, many cultivars could be stored for several months until dried, when food value increases to approximately 335 calories/100 g (Watt and Merrill 1963). Frequently, for longer storage, particularly for thick-fleshed varieties, the fruits were cut into strips and dried in the sun or by some other sufficient method.

Within the Central Plains subarea, almost all cucurbit remains are seeds. Only a few rind fragments have been described from two Steed-Kisker sites (Table 3.1). The earliest known cucurbit cultivation is associated with the Middle Woodland Hopewell occupation at the Trowbridge site (14WY1). Located in features with cultivated marshelder, *C. pepo* seeds are well dated to A.D. 250 (A. Johnson 1979:86). Unlike those from earlier occupations in the Missouri Ozarks, the Trowbridge squash seed measures only 7.8 × 4.6 mm (no allowances for charring). It is unclear at this time whether this size represents immature fruits, a smaller seed variety than the Phillips Spring sample or a smaller than average seed. The Phillip Springs seeds average 9.7 mm in length and 6.5 mm in width (Kay et al. 1980).

A long period of time elapsed before squash appears again in an archaeological context in the Central Plains (Table 3.1). Charred seeds from the Two Deer site (14BU55) in southcentral Kansas were recovered from a house interior with a weighted date of A.D. 1000 ± 25 (Adair and Brown 1981). Although not extremely abundant (.7% of identified seeds), these seeds were similar in size to the Trowbridge sample, averaging 7.9 mm in length and 4.4 mm in width (Table A.1). From approximately the same time period, pepo squash and bottle gourd seeds were recovered from the Walker Gilmore site in eastern Nebraska (Fig. 3.1). Recent excavations at this site have produced new dates of A.D. 1116 ± 63 and A.D. 1174 ± 76 from the cultigen bearing occupations (Haas 1983). While the fairly large sample (about 700 *C. pepo* seeds) was recovered from four Woodland occupation levels, the overall size of the squash seeds appears to be consistent throughout. The charred seeds range in size from 14.7-8.4 mm in length and 10.9-5.4 mm in width. This variation in seed size has been attributed to the presence of several varieties of *C. pepo* in the collection (Cutler and Blake 1983).

A Middle Ceramic Pomona occupation in eastern Kansas, dated to A.D. 875 ± 75, has also produced evidence of squash cultivation. Three charred *C. pepo* seeds have recently been identified from Hatcher (14DO19) and average in size to 7.7 × 4.2 mm.

Remains of this species are also known from Steed-Kisker sites in the Kansas City area, including the Steed-Kisker site (23PL13), Friend and Foe (23CL113), and 23PL16. Sizes of seeds from the latter two sites average 7.6 × 4.5 mm and 9.4 × 5.0 mm, respectively.

Radiocarbon dates place the occupations around A.D. 1000-1200 (Table 2.2). Remains of bottle gourd, including seeds and rind fragments were also recovered from 23CL115 and 23PL16 (Table 3.1).

Further west of the Steed-Kisker area, cultivated squash was recovered from the Upper Republican component at 14ML16 (Cutler and Blake 1973b:31). While little information is available concerning the size of the *C. pepo* seeds, the site is radiocarbon dated to A.D. 1190 ± 90 thus making it contemporaneous with the Steed-Kisker occupation.

Pepo squash seeds have also been recovered from 13ML129 and 13ML138, two Glenwood occupation lodges. Twenty-one measurable specimens from the first site average 7.8 × 4.2 in size with a length/width ratio of .53. Over 100 seeds were identified from 13ML138 (Tables 3.1, A.1). They are comparable in size to the first sample, averaging 7.5 × 4.0 and having a length/width ratio of .53.

As illustrated in Figure 3.1 and Table 3.1, neither squash nor gourd are particularly abundant from sites in the Central Plains prior to about A.D. 1000. After this time, squash remains become a common subsistence item from Central Plains and Middle Missouri sites. The absence of cucurbit remains prior to this time, particularly during the Archaic period, is somewhat puzzling. Since cucurbit rinds and seeds have been recovered from early levels at Koster dated to 7000 B.P. (Asch and Asch 1982) and from Phillips Spring at 4300 B.P. (Kay et al. 1980), it seems reasonable to assume that cucurbits could have been available for cultivation in the Central Plains during the Archaic, especially since the Late Archaic Nebo Hill phase has been demonstrated to have an eastern orientation and may have been part of a regional exchange system (Reid 1983). However, analyses of floral remains from the Coffey site (Schmits 1978) and the Nebo Hill site (Root 1979), both Archaic occupations in the Central Plains, have indicated instead a subsistence well adapted to the local vegetation. Even with such a subsistence base, squash could have been added without a noticeable change in the economy. The lack of cucurbit remains in Archaic period sites in the Central Plains may be due in large part to preservation and identification problems in conjunction with the fact that few sites of this period have been excavated. Pepo flesh is not conducive to long preservation unless it has first been dried and the seeds of both squash and gourd are not as likely to survive the charring and recovery process as are sunflower, corn, and beans. (King 1985, Yarnell 1964).

The varieties of *C. pepo* produce seeds of different sizes and shapes. The *C. pepo* seeds from Middle Woodland, Plains Woodland, and Middle Ceramic sites in the Central Plains are of particular interest when compared to the samples from the Phillips Spring and Walker Gilmore sites and to the cucurbits grown by Indians of the Missouri River region during the protohistoric and historic periods. According to Cutler and Whitaker (1961:Fig. 3), there are three varieties of *C. pepo*. The first, the common Mandan cultivar, produces seeds that average approximately 14.0 × 9.6 mm with a length/width ratio of .68. Seeds of the Red Lodge variety, also a cultivar of the North Central Plains in historic times, average 15.2 × 8.0 mm and have a length/width ratio of .52. The third variety, *C. pepo* var *ovifera*, is actually a small yellow flowered gourd grown today as ornaments. Seeds of this variety measure approximately 9.1 × 5.6 mm and have a length/width ratio of .62. Gilmore (1977: 64-67) states that the Omaha grew, or at least distinguished between eight different kinds of squash and pumpkin which included *C. pepo* and *C. maxima*. It does not seem unreasonable to assume that they were growing all three varieties of *C. pepo* mentioned above. This assumption is supported by evidence supplied by Cutler and Blake (1973b) on protohistoric and historic Arikara. Seed measurements of *C. pepo* from three sites in South Dakota range from 8.0 × 6.5 mm and 16.5 × 9.5 mm (39CA6), to 18.5 × 9.0 mm (39SL13), to 19.2 × 10.4 mm (39SL19), with 1/w ratios of .81 and .57, .48, and .54, respectively. These ratio figures include the measurements of all three varieties of *C. pepo*. Additionally, Cutler and Blake report both the Red Lodge and Mandan varieties from protohistoric Arikara sites (1973b:62) and *C. pepo* var *ovifera* from both protohistoric Arikara and Initial Coalescent sites (1973b:66).

Table 3.1. Distribution of *Cucurbita* within the Central Plains subarea and surrounding regions.

Archaeological site (with map reference number)	*Sample Size*	*Cultural Affiliation*
1. 23HI216 Phillips Spring[a]	149 squash seeds 15 gourd seeds	Archaic
	3 squash seeds 1 gourd seed	Woodland
2. 23BE146 Boney Spring[b]	5 squash seeds	Woodland
3. 14WY1 Trowbridge[e]	1 squash seed	Kansas City Hopewell
4. 23PL13 Steed-Kisker[c]	2 squash rind fragments	Steed-Kisker
5. 23CL113 Friend and Foe[d]	1 squash seed	Steed-Kisker
6. 14DO19 Hatcher[f]	3 squash seeds	Pomona
7. 14BU55 Two Deer[f]	31 squash seeds	Bemis Creek
8. 25CC28 Walker Gilmore[g]	700+ squash and gourd seeds	Sterns Creek
9. 14ML16[h]	?	Solomon River
10. 23CL115[i]	1 gourd rind fragment	Steed-Kisker
11. 23PL16[i]	2 squash seeds 3 gourd seeds	Steed-Kisker
12. 13ML129[j]	21 squash seeds	Glenwood
13. 13ML138[j]	114 squash seeds	Glenwood

[a] Kay et al. 1980; [b] King and MacMillan 1975; [c] Wedel 1943; [d] Calabrese 1969; [e] Johnson 1975; [f] Adair, this study; [g] Haas 1983; [h] Cutler and Blake 1973b; [i] Blake 1974, 1980; [j] Nickel, n.d.

Pepo seeds from the Phillips Spring site compare favorably in terms of length/width ratio with the common Mandan variety (Kay et al. 1980; King 1985). While the Archaic sample seeds are smaller than the ones grown historically, they have a similar 1/w ratio of .67 and .68. This similarity led the authors to suggest that the pepo squash introduced into the Eastern plains about 4000 years ago gave rise to the Mandan variety grown historically (Kay et al. 1980). When the seed shapes of the Phillips Spring and Mandan variety samples are compared to the ones from the Central Plains subarea (Fig. 3.2), several interesting factors can be observed. First, there is a significant difference in seed shape between the Phillips Spring, Two Deer, Hatcher, 13ML129, 23PL16, and 13ML138 samples, with the size of the samples from the latter five sites more closely resembling the Red Lodge variety (1/w ratios of .54, .54, .53, .53, and .53, respectively; Table A.1). Second, the small samples from the Trowbridge and Friend and Foe sites also do not compare in seed shape to either the Phillips Spring or the historic Mandan. Instead, their shape falls somewhere in between that of *C. pepo* var Red Lodge and *C. pepo* var *ovifera* (1/w ratios of .58 and .59, respectively). Since each of these samples are only represented by one charred seed, however, the range in seed size is unknown, rendering it difficult to determine which of the two varieties they most closely resemble. Nevertheless, given the available data, it appears that the squash introduced and grown in the Central Plains subarea from approximately A.D. 200-1200 was a variety that produced seeds smaller than the variety which eventually gave rise to the historic Mandan. Additionally, the samples from the Two Deer,

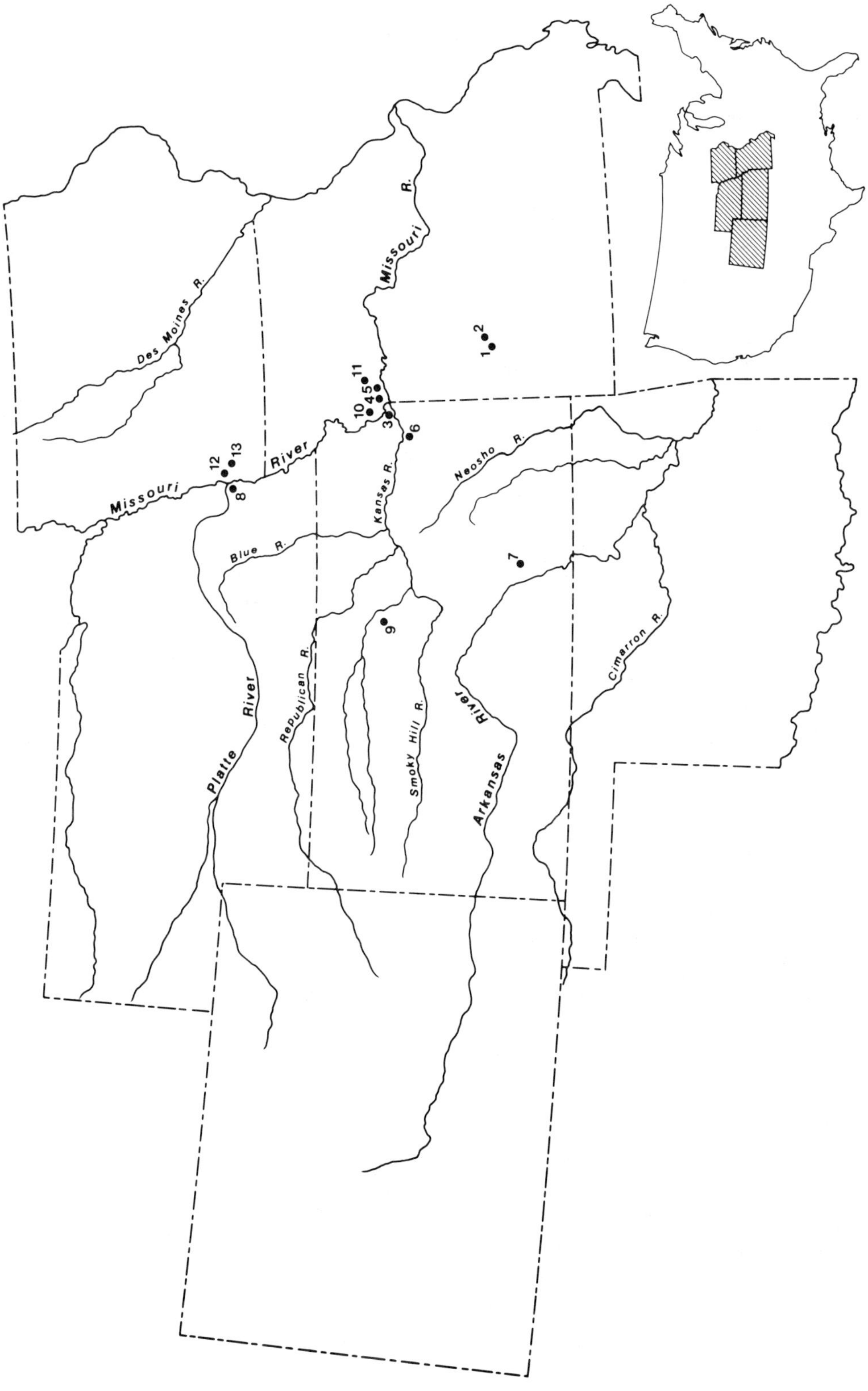

Fig. 3.1 Location of archaeological sites from which cucurbits have been recovered. Numbers refer to sites referenced in Table 3.1.

Hatcher, 23PL16, 13ML129, and 13ML138 sites would suggest that this small seeded squash may have been a predecessor to the Red Lodge variety. While it is possible that the smaller seed size from these sites may have been related to such things as poor growing conditions or immaturity of the fruit when harvested (Richard Ford, personal communication), the shape would have remained the same. Indeed, this is one conclusion made by Kay et al. (1980) when comparing the squash seeds from Phillips Spring to the historic Mandan variety. Seed shape may be a more conservative genetic trait than seed size, indicating that seeds merely enlarged through time without altering shape.

Data from the Walker Gilmore site (13CC28) suggest that by the middle of the 12th century A.D., the Red Lodge variety continued to be grown although a larger seeded variety was of greater importance (Fig. 3.2). This may indicate the introduction of a new variety of pepo squash into the Central Plains sometime between A.D. 1000-1200; an introduction possibly from the East. An increase in seed size may also indicate that the seeds were used for food and the larger seeds were selected for planting. Such a selection for a certain seed morphology could suggest the cultivation of squash had become more advantageous in the Central Plains and that the product was a larger, fleshier vegetable with fewer but larger seeds.

Iva annua *var* macrocarpa *Jackson (marshelder)*

This annual plant, known as marshelder or sumpweed, is one of several native species domesticated prehistorically in eastern North America and the Central Plains. *Iva annua* is native from southern Illinois southward to Mississippi and northeast Mexico and westward along river valleys into the Plains (Jackson 1960; Yarnell 1979). Its ideal habitat is alluvial soils along streams, borders of ponds and sloughs, river bottom prairies and meadows, and low fields in valleys (Steyermark 1963). It is not likely to be found however in all zones that supply a moist, shady habitat but rather exists as an edge species occurring between permanently moist and better drained soils (Asch and Asch 1978:309). Its distribution is linear and generally follows topographic drainage contours where it exists in stands, the densities of which are influenced by the amount of available light (Asch and Asch 1978:309). Occasionally, *Iva* is found growing amongst short grasses in fields frequently grazed or mowed. *Iva*'s ability to grow in soils drier than those found in floodplains suggests that the plant's habitat is controlled in large part by its mode of seed dispersal into floodplain waters (Asch and Asch 1978:319) rather than by its ability to flourish in upland zones.

A second species, *Iva xanthifolia*, is also native to the Central Plains, but its distribution is restricted to the drier portions of western and northern Kansas (Bare 1979; Stevens 1961). Presently, *I. annua* has been found growing in small but dense stands along streams tributary to such major waterways as the Arkansas River, the Missouri River and the Little Blue River in southern Jackson County, Missouri.

Prehistoric use and cultivation of this plant resulted in the development and recognition of a new variety, *Iva annua* var *macrocarpa* (Blake 1939), whose achene size ranged from approximately 4.8 to 10.0 mm in length (Yarnell 1972:336). An increase in seed size presumably resulted from a combination of biological and human actions, an important one being the selection of the largest achenes

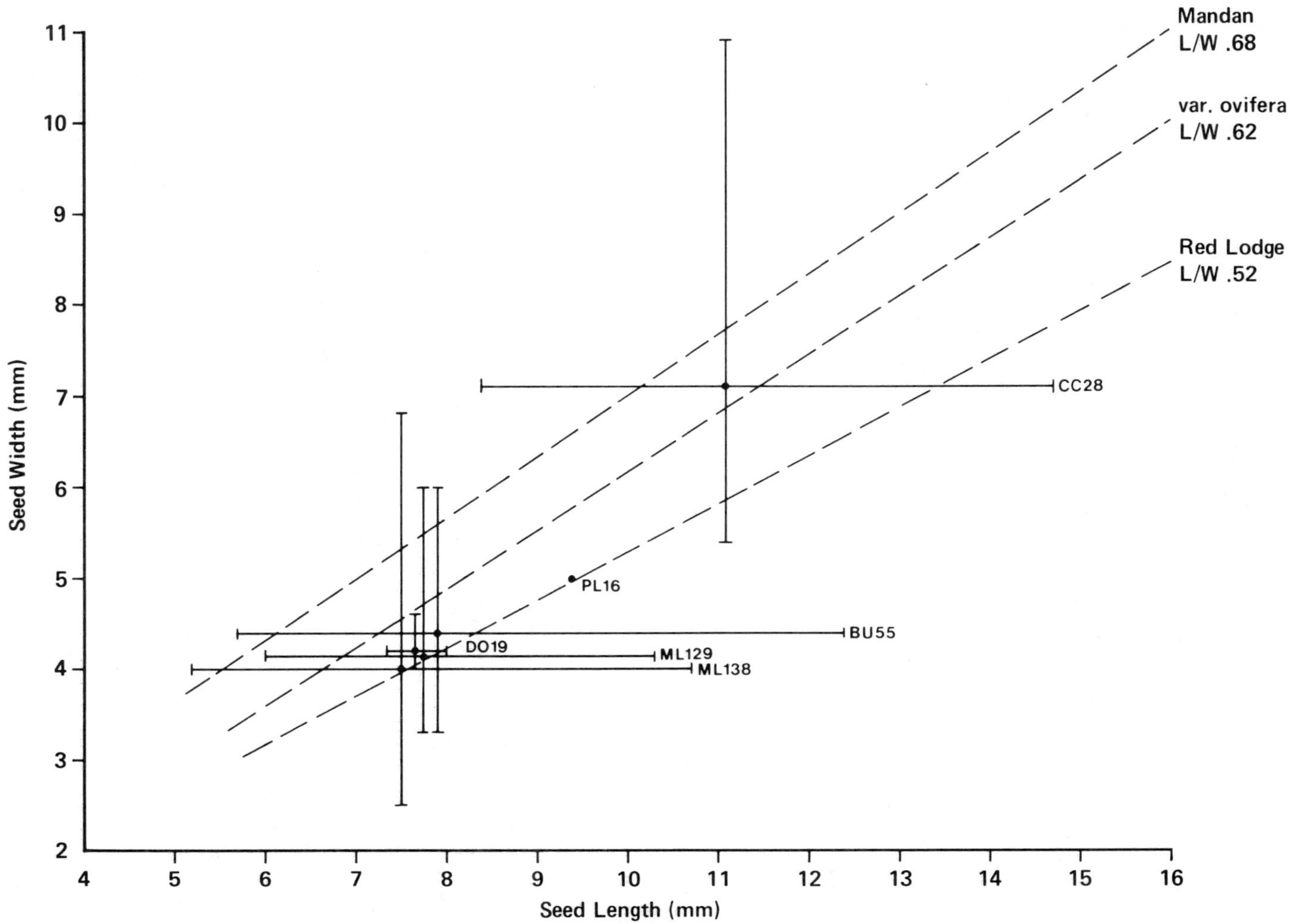

Fig. 3.2 A comparison of squash seed sizes discussed in text. Broken lines refer to the length/width ratios of the three varieties of historic squash. Archaeological samples are represented by solid lines showing ranges in both length and width with the means at the intersection of these lines. CC28 (25CC28 - Walker Gilmore); PL16 (23PL16); BU55 (14BU55 - Two Deer); DO19 (14DO19 - Hatcher); ML129 (13ML129); ML138 (13ML138). Historic squash sizes based on Cutler and Whitaker (1961), figure designed after Kay et al. (1980).

from each harvest for planting the following year. It has also been suggested that *Iva annua* was grown in areas other than its natural habitat (Asch et al. 1972), thereby reducing the amount of introgression between wild and cultivated plants. Such actions would result in a significant increase in seed size in just a few generations (Ronald McGregor, personal communication). Today, *Iva annua* var *macrocarpa* is extinct, as measurements of modern *Iva* achenes indicate they rarely exceed 4.5 mm in length (Fig. 3.3). Steyermark (1963:1536) lists the modern range of *Iva annua* achenes as 2.3-3.8 mm; Asch and Asch (1978:323) suggest that they generally measure from 1.6-4.6 mm; and Jackson (1960) describes them as ranging from 2.0-4.5 mm in length. Samples collected and measured for this study displayed mean lengths of 2.8 to 3.4 mm (Table 3.2). Apparently, there is a range in the sizes of modern *Iva annua* achenes, a factor perhaps related to different growing conditions and date of flowering (Asch and Asch 1978). However, sizes of achenes in excess of 5.0 mm or similar in size to several archaeological samples are unknown from modern *Iva* collections. Thus, the argument in favor of aboriginal domestication of *Iva* is entirely dependent upon the archaeological evidence (Yarnell 1972, 1977). The large sizes of *Iva* achenes, plus their presence in archaeological sites which may lie beyond the modern range of the species are the critical factors supporting *Iva*'s status as a cultivated plant and a domesticate.

Iva plants produce an oily seeded achene, which like that of the sunflower, consists of a single seed or kernel enclosed within a thin dry shell or pericarp. In its unprocessed state, the *Iva* achenes are unpalatable due to the unpleasant odor and taste of the pericarp. Boiling, roasting and rubbing have been found to be effective means of separating the seed from the pericarp (Asch and Asch 1978:302). The remaining *Iva* seed has been shown to be a concentrated source of food energy, rich in protein and unusually high in several essential vitamins and minerals. They provide approximately 535 calories/100 g, supply essential protein requirements and contain high levels of calcium, iron, phosphorus, potassium, riboflavin, and niacin (Asch and Asch 1978:303-308). In fact, few plants compare with *Iva* in nutritional quality; sunflower seeds and nuts being the most similar.

In this study, domesticated varities of native species were determined by a comparison of size of the archaeological remain to modern wild samples. Archaeological samples statistically larger than modern uncultivated samples and which fell within the accepted range of size for cultigen varieties were considered domesticated. In an earlier publication (Adair 1984a), the size correction equation used to reconstruct prehistoric *Iva* achenes and kernels was that suggested by Yarnell (1972:336-337). This correction method has recently been reviewed (Asch and Asch 1978, 1985) with the determination

Table 3.2. *Iva annua* achene sizes from modern collections.

Location	*N*	*Mean*	*standard dev.*	*range*
Kansas:				
Douglas County	24	3.3×2.5	.4×.2	2.7-4.2×2.0-3.1
Anderson County	60	3.4×2.7	.4×.46	2.6-4.3×1.9-3.9
Bourbon County	59	2.8×2.1	.35×.27	2.4-3.7×1.7-2.9
Atchison County	64	3.2×2.4	.4×.4	2.5-4.0×1.6-3.2
Greenwood County	67	3.3×2.6	.45×.3	2.6-4.3×2.0-3.2
Missouri:				
Jackson County	70	3.2×2.5	.4×.2	2.4-4.3×1.9-3.4

that it "overcompensates for pericarp loss in specimens of average wild size and undercompensates for larger specimens" (Asch and Asch 1985: 162). Hence a new correction equation was devised (Yarnell 1985, Asch and Asch 1985) and all measurements changed for this publication. The following equations were used to reconstruct uncarbonized achene dimensions from the observed dimensions of carbonized kernels and achenes:

Uncarbonized achene dimensions from carbonized achenes:

(carbonized achene length) ÷ 0.9

(carbonized achene width) ÷ 0.9

Uncarbonized achene dimensions from carbonized kernels

(carbonized kernel length × 1.36) + 0.17mm.

(carbonized kernel width × 1.45) − 0.06 mm.

In several archaeological samples both carbonized kernels and achenes were present (Table A.2). A mean size for reconstructed samples was obtained by determining averages for reconstructed carbonized kernels and achenes separately and averaging the two results (Yarnell, personal communication).

Iva annua achenes and kernels are not abundant in botanical collections from Archaic period sites in the study area (Fig. 3.3). Despite extensive flotation of matrix from several sites, *Iva* appears in the archaeological record only once during this period. One charred achene was recovered from the Nebo Hill site and after corrections were made for carbonization, the reconstructed achene measured 3.9 mm in length (Root 1979: 243).

The first definite evidence of *Iva* domestication comes from the Kansas City Hopewell Trowbridge site where 15 *Iva annua* kernels and achenes were recovered from two trash-filled storage pits. Most of these remains were located in Area B, feature 3 (E. Johnson 1972), the same feature that produced the squash seed. Charcoal from this feature provided radiocarbon dates of around A.D. 250 (Table 2.2). The reconstructed mean achene size is 6.4 × 4.1 mm (Table A.2)

Iva cultivation may not have been a practice common to all Kansas City Hopewell sites. Charred *Iva* kernels were also recovered from the Young site, a large village located on a terrace of Brush Creek, a tributary of the Missouri River. The contents of seven features contained marshelder seeds along with a variety of other native flora (Adair 1977). Dense stands of *Iva annua* were observed in the fall of 1982 growing along the margins of Line Creek, a tributary located just east of Brush Creek and also flowing into the Missouri River. If prehistoric vegetation patterns were similar to those observed today, the presence of this plant suggests that it was available to prehistoric populations of the Kansas City region. However, the charred seeds from the Young site do not appear to be of the cultigen variety. After corrections for charring were made, the reconstructed achenes are almost identical in size to a modern fresh sample (Table 3.3), with mean measurements of 3.3 × 2.3 mm (Table A.2). However, a single kernel recovered from feature 40 of this site is strikingly different in size, measuring 7.9 × 5.0 mm. Such a size is well within the range of the domestic variety. Additionally, charcoal from this feature produced a date of A.D. 610+60-75, suggesting that *Iva* domestication was adopted late in the occupation span at this site, domestication and collection of natural stands of *Iva* occurred simultaneously, or that the kernel belongs to an occupation that post-dates the Middle Woodland.

Additional *Iva* domestication during the Middle Woodland period is documented by the recovery of one kernel from the Elliott site in northcentral Kansas (O'Brien et al. 1973). Previously unpublished, the reconstructed size of this kernel is 5.6 × 4.7 mm (Table 3.4). The Elliott site is interpreted as a Middle Woodland Schultz phase occupation.

The continued cultivation of *Iva* into the early Late Woodland period is documented by evidence from the Yeo site (O'Brien 1982). Yeo is thought to be a small special purpose Kansas City Hopewell site radiocarbon dated to A.D. 765. Flotation recovered 58 charred marshelder achenes and kernels, most of them fragments. Of this sample, 19 measurable specimens had an average reconstructed length of 5.3 mm and a width of 4.2 mm (King 1982a:3; Asch and Asch 1985:163). Although smaller than the Trowbridge samples, this seed size is still within the accepted range of cultigen *Iva*.

The smaller seed size noted from the Yeo site continues into succeeding periods. Charred *Iva* kernels recovered from the Mississippian Friend and Foe (23CL113) site have an average reconstructed size of 6.6 × 4.7 mm (Table 3.3). The cultigens were all associated with features of House 3 which produced radiocarbon dates of A.D. 1050 and A.D. 1100 (Table 2.2). Additionally, marshelder was recovered from features at sites 23CL274 and 23CL276 (McHugh 1982). A single achene from the latter site measures 4.8 × 2.4 mm while site 23CL274 yielded eight charred kernels averaging 5.9 × 3.6 mm in size (Table 3.3).

Excavations from the Glenwood locality in southwest Iowa (Fig. 3.3) provide additional information on *Iva* domestication. Of 11 sites containing botanical remains of domesticated plants, three yielded charred *Iva* achenes and kernels; some in sizeable quantity (Table 3.3). Reconstructed measurements for 392 specimens from 13ML129 were 6.1 × 4.4 mm. Twelve kernels and achenes from site 13ML138 were reconstructed to an average of 6.0 × 3.7 mm and a single achene from site 13ML139 was recon-

Table 3.3. Distribution of *Iva annua* within the Central Plains subarea.

Archaeological Site (with map reference number)	*N*	*Mean*[a]	*Range*
1. 23CL11 Nebo Hill[b]	1	3.9 × ?	-
2. 23PL4 Young	19*	3.3 × 2.3 = 8	2.3–7.9 × 1.7–5.0
3. 23CL199 Yeo[c]	19	5.3 × 4.2 = 22	3.7–6.8 × 2.8–5.3
4. 23CL113 Friend and Foe[d]	9	6.6 × 4.7 = 31	5.7–7.8 × 3.7–5.5
5. 23CL274[e]	8	5.9 × 3.6 = 21	5.5–6.6 × 2.9–4.0
6. 23CL276[e]	1	4.8 × 2.4 = 12	-
7. 23CL115[f]	17	5.7 × 4.0 = 23	4.5–6.8 × 3.4–4.9
8. 23PL4 Young[f]	4	5.9 × 3.7 = 22	4.8–6.7 × 3.1–4.2
9. 14WY1 Trowbridge	15*	6.4 × 4.1 = 25	5.5–7.7 × 2.8–6.0
10. 23SA128 Fisher-Gabert	2*	6.3 × 4.0 = 25	-
11. 13ML129[g]	392*	6.1 × 4.4 = 27	3.8–9.5 × 1.9–6.3
12. 13ML138[g]	12*	6.0 × 3.7 = 22	4.5–7.9 × 1.9–4.7
13. 13ML139[g]	1*	5.5 × 4.2 = 23	-
14. 14BU55 Two Deer	21*	5.7 × 3.9 = 22	4.3–7.4 × 3.3–5.0
15. 34WN29 Drumming Sauna	1*	5.7 × 4.9 = 28	-
16. 14ML371[h]	?	?	-
17. 14ML417	5*	6.0 × 3.9 = 23	5.2–6.4 × 3.3–4.6
18. 14GE600 Witt	8*	6.3 × 4.3 = 27	4.9–7.8 × 3.6–5.0
19. 23JA269 Vaughn Estess[i]	3	6.7 × 5.0 = 34	6.2–7.3 × 4.3–5.7

[a] size corrected according to Asch and Asch 1985; [b] Root 1979; [c] King 1982a; [d] Calabrese 1969; [e] McHugh 1982; [f] Blake 1974, 1980; [g] Nickel n.d.; [h] Cutler and Blaker 1973b; [i] Voigt 1985
* measured by author
? refers to unknown amounts or size—not included in report

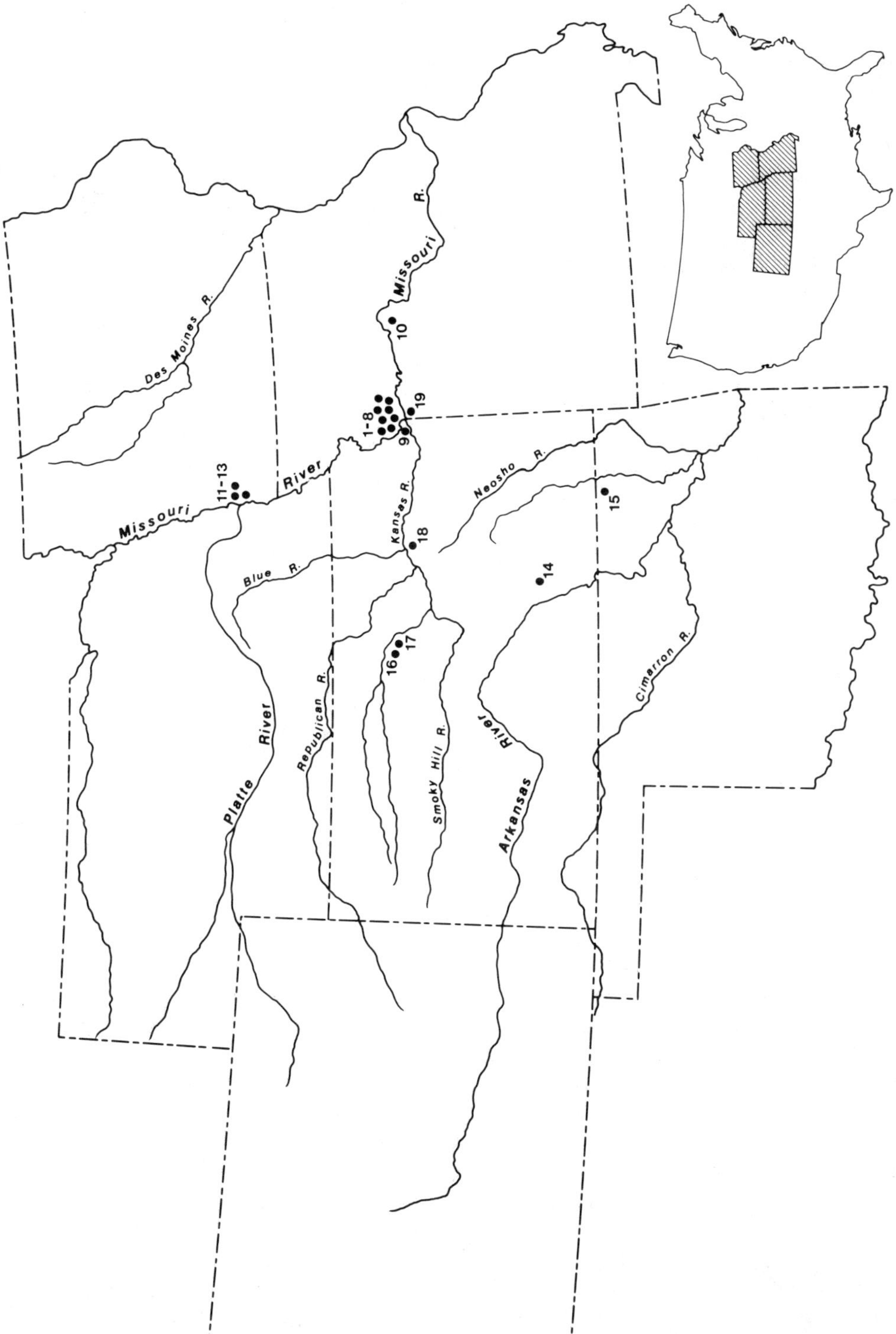

Fig. 3.3 Location of archaeological sites from which *Iva annua* has been recovered. Numbers refer to sites referenced in Table 3.3.

structed to 5.5 × 4.2 mm (Tables 3.3, A.2). Radiocarbon dates from these sites cluster within the 12th century A.D. (Table 2.2).

Additional evidence for *Iva* domestication at A.D. 1000 comes from the Two Deer site in southcentral Kansas (Fig. 3.3). Twenty-one of the 26 recovered charred achenes and kernels were complete enough to measure, providing a mean reconstructed size of 5.7 × 3.9 (Table 3.3). The *Iva* remains from Two Deer are small in comparison to the earlier Middle Woodland samples, but when compared to five modern fresh samples collected throughout eastern Kansas, they are significantly larger in both length and width (Table A.2).

Further evidence for *Iva* domestication is found in the Central Plains tradition. The Upper Republican Solomon River component at 14ML417 yielded meager data on plant exploitation; however five charred *Iva* kernels were extracted from a single feature. The reconstructed achene length ranges in size from 5.2 to 6.4 mm and averages at 6.0 mm (Table 3.3). The importance of this plant in the diet is somewhat questionable. Since all of the seeds were recovered from the same feature, they could easily represent a single plant, suggesting that the marshelder was of less value in the diet than other plant foods. However, the sizes of the achenes are statistically larger than fresh modern samples, indicating at least enough human involvement to produce a cultigen variety.

Cultigen *Iva* was recovered from two additional Middle Ceramic sites in the study area. The Smoky Hill occupation at the Witt site (14GE600) yielded eight kernels and achenes which display mean reconstructed measurements of 6.3 × 4.3 mm. A recent excavation at the Vaughn Estess site (23JA269) provided the first evidence of *Iva* domestication from a May Brook occupation (Voigt 1985). Three kernels were reconstructed to an average size of 6.7 × 5.0 mm (Table 3.3)

To the southern part of the study area, within northeastern Oklahoma, is the Drumming Sauna site (34WN29). Here a single kernel was identified with a reconstructed size of 5.7 × 4.9 mm. The occupation at this site is interpreted as the winter base camp of a small group of aceramic Caddoan hunter-gatherers who revisited the site on several occasions (Reid and Artz 1984). Radiocarbon dates for this occupation are A.D. 740 ± 110 and A.D. 990 ± 50.

One would expect the size of *Iva* achenes to increase through time if the plant continued to be domesticated. Such an increase has been observed from archaeological samples dated from the Late Archaic to the Mississippian period in the Eastern United States (Yarnell 1981). The product of mean length times mean width in mm (mean achene size index) for sizeable collections of *Iva* suggests a continuing, and somewhat uniform, increase in size from 13 to 21 to 35 for three broad periods; Terminal Archaic and Early Woodland, Middle Woodland and Early Late Woodland, and Mississippian (Yarnell 1972, 1979, 1981). Such a trend is not noticeable in *Iva* remains from Central Plains archaeological deposits. Instead, the mean achene size index (L × W) appears to remain between 22-28 mm from the Middle Woodland to the Middle Ceramic periods (Table 3.3). This may suggest that cultigen *Iva* remained at a relatively constant level of importance for an extended period of time despite the domestication of other crops. It may also suggest, as Cowan (1985:209-210) points out, that prehistoric gardeners found it difficult to improve marshelder to the point where a homogeneous crop could be harvested, and the mean size remained relatively constant as long as cultivated varieties continued to be associated with wild plants. As stated previously, increased seed size is one of the most noticeable effects of cultivation on plant evolution (Baker 1972) and one that resulted from human's deliberate selection of large seeds for planting. For *Iva*, the continuing increase in seed size may have been due to polyploidy and selection in favor of genes for large seed size, rather than to excessive watering and/or fertilizing by man (Yarnell 1981: 297). If the selection process was not maintained year after year, seed size would eventually decrease due to introgression with wild plants and the variety would become extinct. Perhaps this is what happened to *Iva annua* var *macrocarpa*. Much like the beginning of *Iva* cultivation, decline in achene size would also be a process. When this process was complete is not actually known. Europeans made no surviving report of *Iva* domestication in North America while only one 16th century protohistoric site is known to yield cultigen *Iva* (Nickel 1977). Yarnell (1972:340) suggests that marshelder cultivation had stopped in the eastern Woodlands soon after the middle of the second millenium A.D. When *Iva* domestication terminated in the Central Plains is unknown, although data suggests it may have been during the 13th century. Other domesticated crops, particularly the tropical cultigens, increased in importance by this time. It has been suggested that a polycultural crop system may have actually played a part in *Iva*'s decline (Asch and Asch 1978:334). *Iva* is a monocultural plant, growing best and producing the most when it is not surrounded by competing plants, while other crops, especially corns and beans, are quite productive when grown together.

Cowan (1985:210) has recently suggested that it may not be possible to assign isolated *Iva* achenes to a specific cultural/temporal period solely on the basis of size, since size distributions may become more widely dispersed through time. However, achene size does increase through time, perhaps on the order of 1 mm/1000 years (Yarnell 1978). Within the Central Plains, data are practically nonexistent to demonstrate a long period of *Iva* utilization that culminated in its domestication. It first appears in measurable quantity during the Middle Woodland period and is definitely identified as the domesticated variety. While the origins of the Middle Woodland Kansas City Hopewell are not clearly known (A. Johnson 1979:86-87), a long term social interaction with the Illinois Valley Middle Woodland populations has been demonstrated. Similarities in subsistance have been noted, and may include the domestication of the native annual *Iva annua*. The longest documented use of this plant, a use that eventually led to its cultivation and domestication, comes from the lower Illinois River Valley (Asch and Asch 1982, 1985). Research in this area has provided documentation of the use of this plant from an Archaic level occupation at the Koster site dated to 7300 B.P. Charred achenes from a Titterington phase pit, dated to 3920 B.P. at the Napoleon Hollow site measure 4.2 mm in length and may provide the earliest evidence for *Iva* domestication (Asch and Asch 1982:6). By the Middle Woodland period, *Iva* domestication is well documented from many sites with achene length measurements averaging from 4.8 to 6.2 mm (Yarnell 1979, 1981). A similar domestication process is not known for any region in the Central Plains.

While an indigenous Early Woodland complex in the Kansas City area may have aided in the development of the Kansas City Hopewell (A. Johnson 1979; Wright 1980), macrobotanical samples from two of these Early Woodland sites, 23JA36 and 23JA159 (Table 2.3), do not contain remains of *Iva* and only one *Iva* achene is known from Archaic deposits. The earliest occurrence of this plant in any abundance in an archaeological context is from the Trowbridge site, where achene size shows it to be from the domesticated variety. Interestingly, the corrected measurements of the charred achenes fall within the range of that observed for the Illinois River Valley Middle Woodland samples. Not only are the means in agreement, but the range of sizes as well (Table 3.4), perhaps suggesting a parallel development of *Iva* domestication. Given the amount of contact between the Kansas City Hopewell and Middle Woodland populations to the east and the fact that no positive evidence of *Iva* utilization is known in the Kansas City region prior to A.D. 200, the occurrence of domesticated marshelder at the Trowbridge site may be related to influence from populations to the east who had a long history of *Iva* utilization and domestication (Asch and Asch 1985). *Iva* domestication may therefore not have developed in the Kansas City area but was introduced from the east. Additional support for this suggestion comes from the Fisher-Gabert site, a Middle Woodland occupation in central Missouri. *Iva* achenes were recovered from features whose contents included Havana and Hopewell series vessels (Kay and Johnson 1977) and are dated from approximately A.D. 200-400. The charred cultigens had reconstructed sizes well within the range of the domestic variety (Table 3.4).

Helianthus annuus *var* macrocarpus Ck11. (sunflower)

The sunflower is a genus containing both annual and perennial species, several of which are common to the Central Plains and easily identified by their distinctive yellow flower. The plant is noted for its hardiness, its high potential productivity, and its ability to thrive in disturbed soil situations. The common sunflower (*Helianthus annuus*) is often one of the first plants to inhabit a disturbed area and usually dominates until eliminated by developing ecological succession (Heiser 1976; Yarnell 1979). It is likely to be found in a variety of ecological zones but grows best in mesic to moderately xeric soil moisture conditions.

The history of the common sunflower, which includes the distribution of the various uncultivated species as well as the emergence of the domesticated form, is described by several interpretations. According to Anderson (1956), prolonged introgression of the common sunflower (*Helianthus annuus*) and the prairie sunflower (*H. petiolaris*) within the Colorado Plateau area produced the

Table 3.4. A comparison of *Iva* lengths between Kansas City and Illinois Valley Hopewell.

Site	*N*	*Mean L × W = LW*	*Range*
Macoupin[a]	193	5.3 × 3.7 = 20	2.9–9.8 × 1.7–6.5
Apple Creek[a]	19	5.7 × 4.2 = 22	4.3–6.8 × 2.6–5.7
Stillwell[a]	65	6.2 × 4.1 = 25	4.8–7.5 × 2.9–5.8
Newbridge[b],[e]	229	6.0 × 4.3 = 26	3.2–9.3 × 2.1–6.7
Trowbridge[c]	15	6.4 × 4.1 = 25	5.5–7.7 × 2.8–6.0
Yeo[d]	19	5.3 × 4.2 = 22	3.7–6.8 × 2.8–5.3
Fisher-Gabert[c]	2	6.3 × 4.0 = 25	–
Elliott 14GE312[c]	1	5.6 × 4.7 = 27	–

[a] Yarnell 1979; [b] Asch and Asch 1982; [c] Adair, this study; [d] King 1982a; [e] Yarnell, personal communication

Great Plains race of *H. annuus* which spread eastward through the prairies. From the Great Plains *H. annuus* came both the cultivated variety and the weedy campfollower variety of the sunflower. Although Heiser (1985:57-58) documents the greatest concentration of the annual species of *Helianthus* to be in the Southwest, he suggests that the larger weedy form of sunflower found in the central and eastern United States probably originated from the introduction of the wild sunflower from the West (Heiser et al. 1969). The weedy sunflowers found in these areas today do not appear to be escapes from cultivation (Heiser 1985:58) although some hybridization with the domesticated variety may occur.

Based on measurements provided by Heiser (1954, 1978), the original Great Plains *annuus* probably had achene lengths of 4.5 to 5.0 mm. Yarnell (1979:291) states that modern ruderal sunflower achene lengths range from 4.0 to 7.0 mm. Modern *Helianthus annuus* achenes collected from Butler and Sumner counties in Kansas had mean lengths of 3.9 to 5.2 mm, and fall in the lower range of sizes reported by Heiser (1954, 1978). It seems reasonable, though, to expect some variation in size between Central Plains *annuus* achenes and those of the eastern Woodlands. In addition, it appears that *H. annuus* and *H. petiolaris* intergrade quite commonly in the Central Plains today (Heiser et al. 1969:68; Stevens 1961:378), to the point that *H. petiolaris* should perhaps be regarded as a depauperate form of *H. annuus*. It seems quite possible that the common sunflower found today in the Central Plains produces smaller achenes than the variety to the east due to its constant hybridizing with smaller achene varieties such as *H. petiolaris*. That this introgression is only a modern occurrence cannot be taken for granted, since *H. petiolaris* is native to the western Great Plains (Anderson 1956) and is not known to have spread outside of the Plains in pre-Columbian times. Therefore, the small seeded achene lengths reported above for southcentral Kansas appear to represent wild *Helianthus annuus* and will be compared to sizes of archaeological samples. Measurements provided by Yarnell for ruderal sunflowers will be used in this study. Modern cultivated achenes range in length from about 6 to 20 mm (Steyermark 1963; Yarnell 1977).

The sunflower is similar in many respects to the marshelder. Both are members of the family Compositae and the seeds of both are similar in color, texture, flavor, protein and oil content (Barclay and Earle 1974; Yarnell 1979). Their nutritional values do not differ significantly with the meat of the sunflower seed containing about 52% protein (Steyermark 1963:1569). Additionally, the fruit of each is an achene that consists of a single seed enclosed in a dry, indehiscent pericarp. One major difference however between the sunflower and the marshelder is the ease in which the pericarp can be removed. Marshelder requires greater work while the sunflower can be "shelled" quite easily when the fruit is mature and dry. In addition, the sunflower pericarp is not totally unpalatable. Even today the whole achene is packaged and consumed as a snack.

Following the introduction of cultivated squash and gourd into the Eastern United States, the domestication of native sunflower is recognized in the archaeological record beginning during the early first millennium B.C. Sunflower seeds measuring within the size range of the domesticated variety have been recovered from Salts Cave and Mammoth Cave in Kentucky, the Higgs and Westmorland-Barber sites in eastern Tennessee, and the Riverton site in Illinois (Ford 1981; Yarnell 1976). Increasing size is recognized in achenes from Middle and Early Late Woodland sites in the same geographical areas. Beginning during the latter part of the Late Woodland or early Mississippian period (A.D. 900-1000), sunflower achenes increase significantly in size and appear in sites throughout the eastern Woodlands, the Middle Missouri area, and into the Plains. Yarnell (1981) suggests that since achene collections from the same general time period fall within a relatively restricted mean size range, undated collections can be temporally placed by comparisons. The reliability of using sunflower achene size as a chronological indicator is somewhat weakened by a significant development in varietal diversity (Heiser et al. 1969).

Nineteen sites within the study area have yielded sunflower achenes or seeds. While sample sizes are generally small, several provide excellent documentation for domestication (Figs. 3.5 and 3.6). Measurements were corrected to compensate for shrinkage due to charring and, if necessary, for loss of pericarp. Reconstruction formulas used are as follows (Yarnell, personal communication):

carbonized achene: length ÷ 0.9
width × 1.2
carbonized kernel: length × 1.3
width × 1.45

Six Archaic sunflower achenes recovered from the Coffey site are, to date, the oldest identified *Helianthus* from the Central Plains area. Recovered from Horizon III-3, they have an average reconstructed size of 4.3 × 2.5 mm which is well within the range of modern wild *Helianthus* (Table 3.5 and A.3). Continual utilization during the Early Woodland period has so far not been documented. Identified flora from the Kansas City Hopewell occupation at the Young site includes only one partial *Helianthus* achene (Adair 1977). Although one fragment does not document utilization, evidence from Middle Woodland occupations to the East attest to the use of this plant during this time (Asch and Asch 1982; Yarnell 1981). It does not seem unreasonable to assume that Kansas City Hopwell populations had access to and gathered wild *Helianthus* achenes also.

Additional use of this plant comes from site 14LT304, a Cuesta phase occupation in southeastern Kansas. Many aspects of this phase exhibit similarities to the more northern Middle Woodland occupations and it is possible that similar subsistence strategies were also shared. At present, it is not known whether the one fragment represents a wild or cultivated variety although speculations are that the achene is from a wild variety.

The first strong evidence for *Helianthus* cultivation and domestication comes at about A.D. 900-1000 from several sites in eastern Kansas. Approximately 130 charred achenes and seeds were recovered from the interior of a house at Two Deer (14BU55), and of this amount, 120 were complete enough to measure. Given the correction factors for shrinkage following charring and loss of pericarp, the mean measurements for these achenes are: length - 9.1 mm, width - 4.4 mm, and a length × width product of 40. Figures 3.4 and 3.5 adequately display the size differences between this sample and the size range of modern wild samples.

An additional site of about the same time as Two Deer also provides evidence of sunflower domestication. Hatcher (14DO19), a Middle Ceramic Pomona occupation in northeastern Kansas, yielded eight charred *Helianthus annuus* seeds. Corrected mean measurements for these are: length - 10.1 mm, width - 4.5 mm, and a length × width product of 46. Although sample size is small, they appear to be overall larger than those from Two Deer (Figs. 3.5, 3.6 and Table A.3).

Charred sunflower achenes and seeds from other Middle Ceramic sites provide additional data on the domestication of this annual. Yarnell (1979, 1981) reports that six achenes from the Steed-Kisker site (23PL13) had a mean size of 10.8 × 5.3 mm and a LW product of 58. Flotation of several storage pits associated with the Steed-Kisker occupation at Young (23PL4) produced a larger sample of *Helianthus* and corrected measurements provide a mean size of 8.8 × 3.9 mm (Table 3.5, Table A.3). Additional data on Steed-Kisker cultivation is provided by Wedel (1943). Floral remains from three pits also at 23PL13 were analyzed by Volney Jones in 1940 who described the sunflower achenes as measuring about 9 mm in length. Two other Steed-Kisker sites, 23CL115 and 23CL109, yielded a small sample of sunflower achenes that measured 6.4 × 3.2 mm and 6.8 × 3.6 mm respectively (Table 3.5). Obviously a significant size range in sunflower achenes can be observed for the Steed-Kisker phase, and the reason(s) for this are debatable. To begin with, the sample sizes are small and the ranges in size can only be inferred at this point. According to Yarnell (1976, 1979), however, all samples fall at least within the lower limits of the size range of the domesticated variety while a few fall within the upper limits of the range of the ruderal varicty (Table 3.5). A closer examination of the Young site sample (Table A.3) suggests that almost the entire ranges of these two varieties are present, with sizes ranging from 5.1-12.4 mm in length. Steed-Kisker populations may have been cultivating sunflower in garden plots while harvesting from plants volunteering around the habitation areas. An alternate situation that would produce seeds of such varying sizes would involve the cultivation of sunflower in areas where the plants could easily hybridize with their wild counterparts. Speculative reasons aside, the widely dispersed size ranges of sunflower achenes may simply be a result of the nonuniformity of achene size within a single disk (Heiser 1985). Cultural factors and environmental conditions may be either insignificant to the

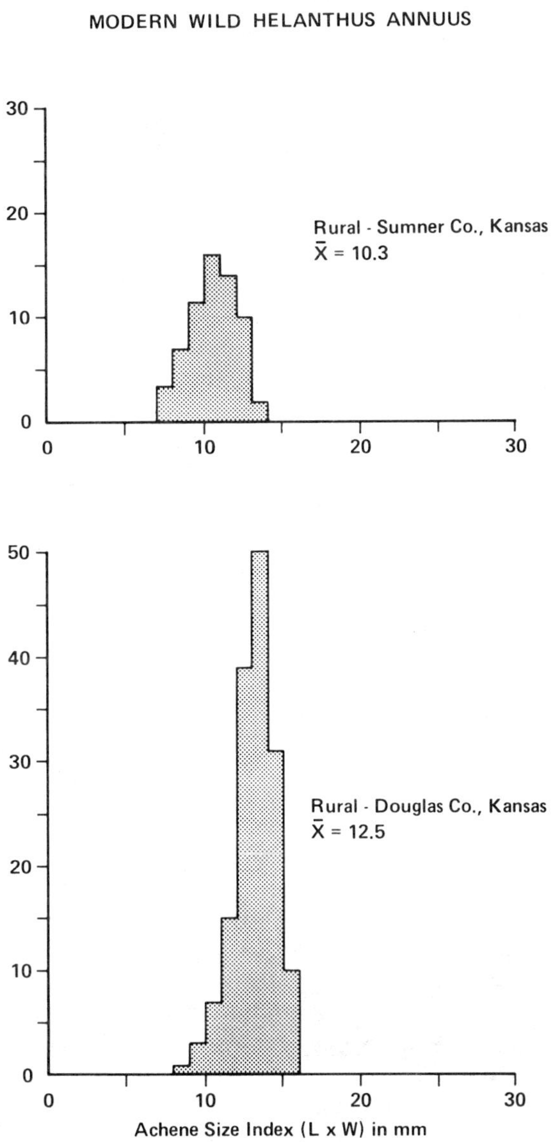

Fig. 3.4 Achene size index (L × W) in mm for modern wild *Helianthus annuus* achenes.

issue or may contribute in yet an undetermined amount.

Whatever the reason, small seeded cultivated sunflower have also been recovered from sites 14BU71, 14ML16, 14ML417 and 14CF508 (Fig. 3.6, Table 3.5). All of these sites are temporally placed within the Middle Ceramic period and radiocarbon dates span A.D. 1200-1400 (Table 2.2). Reconstructed mean lengths of sunflower achenes range from 5.6 to 10.4 mm and may represent a mixing of the common sunflower with the prairie sunflower (*Helianthus petiolaris*).

Three Glenwood locality sites have yielded *Helianthus* achenes and seeds. Over 300 specimens from 13ML129 average 9.2 × 3.8 mm while the mean size of 22 seeds and achenes from 13ML138 is 9.8 × 3.9 mm. A single achene from 13ML139 measures 11.1 × 3.7 mm. These mean reconstructed sizes and size distributions (Table 3.5) are similar to the trends noticeable in the Steed-Kisker collections.

Unlike marshelder, sunflower domestication has continued in the Central Plains during historic times. Wedel (1959) makes numerous references to the presence of sunflower achenes in Central Plains tradition sites. The plant's importance is also well documented from the protohistoric period from sites in the northern Plains area. Will and Spinden's monograph (1906) documents the cultivation of this plant, along with corn, beans, squash, and tobacco, by the Mandan during the eighteenth and nineteenth centuries. Gilmore (1977:78) adds the Arikara and Hidasta to the list of historic populations who cultivated this annual. Several of these accounts describe field conditions in which sunflowers were planted along the edge of gardens, surrounding the corn, beans, and squash. Such a technique may have allowed cultivated sunflower to crossbreed with the wild or ruderal varieties which would have produced *Helianthus* achenes of small and variable sizes

Wilson (1977) describes several varieties of sunflowers cultivated by the Hidasta, including large, big headed plants whose seeds were the earliest to be harvested, and smaller headed varieties known as "baby sunflowers." Additionally, wild sunflowers, similar in size to the baby cultivated varieties, were also gathered. The Hidasta made a distinction between the oil extracted from the different plants, with the smaller headed varieties actually being preferred (Wilson 1977:18-19). Since the seeds from these plants were always gathered late in the season, most were touched by frost which seemed to bring out the oil. The oiliness brought out by the frosting was more apparent in the seeds of the smaller sunflowers than in the seeds of the larger heads

Despite the archaeological and ethnographic accounts attesting to the continual domestication of *Helianthus*, data from the Central Plains are not in agreement with Yarnell's study (1981) suggesting a corresponding achene size increase. In the Eastern Woodlands, cultigens from Early Woodland sites have mean LW products of 22 mm to 25 mm; Middle Woodland samples range 25 mm to 35 mm; Early Late Woodland samples range 35 mm to 60 mm; and Mississippian samples have mean products of 50 mm to 100 mm (Yarnell 1981). With the exceptions of the six achenes measured by Yarnell from the Steed-Kisker site and the one kernel from 14MN509 (a Great Bend occupation), the mean LW products listed in Table 3.5 and displayed in Fig. 3.5 from sites in the Central Plains are consistently smaller than one would expect for each respective period. Complete achenes are not present in Middle Woodland samples, making comparisons impossible.

The Mississippian Steed-Kisker samples range in size from 21 to 58 with Yarnell's sample and 21 to 34 without it. The Two Deer and Hatcher samples (ca A.D. 800-1000) have larger LW means but are still smaller than the Late Woodland samples given by Yarnell (1981). Asch and Asch (1985) have provided data from northeastern Illinois that provide additional documentation of the occurrence of small achenes in late contexts. They suggest that a greater allowance should be made for size variability in samples from late prehistoric times, as this would be in accord with Heiser's (1951) account of size variability within traditional Indian varieties of the twentieth century (Asch and Asch 1985:170).

The similarities between marshelder and sunflower mentioned earlier prompts the suggestion that *Iva* domestication declined in favor of *Helianthus* as a more productive oily-seeded, protein-rich plant. Recognizing that this suggestion is somewhat speculative at this point, there are nonetheless several plausible reasons. First, while both are oily-seeded achenes with similar nutritional values, the archaeological record indicates that *Iva* domestication declined significantly at about the same time as sunflower cultivation began. Second, the sunflower is a hardier plant and more flexible in its adaptation to varying habitats, while *Iva* grows best in damp alluvial soils along stream margins. Third, because the achenes of the sunflower are larger and more nucleated than those of *Iva*, the sunflower has an obvious advantage in attracting greater attention. Fourth, the sunflower achene is available for harvesting beginning about mid-September

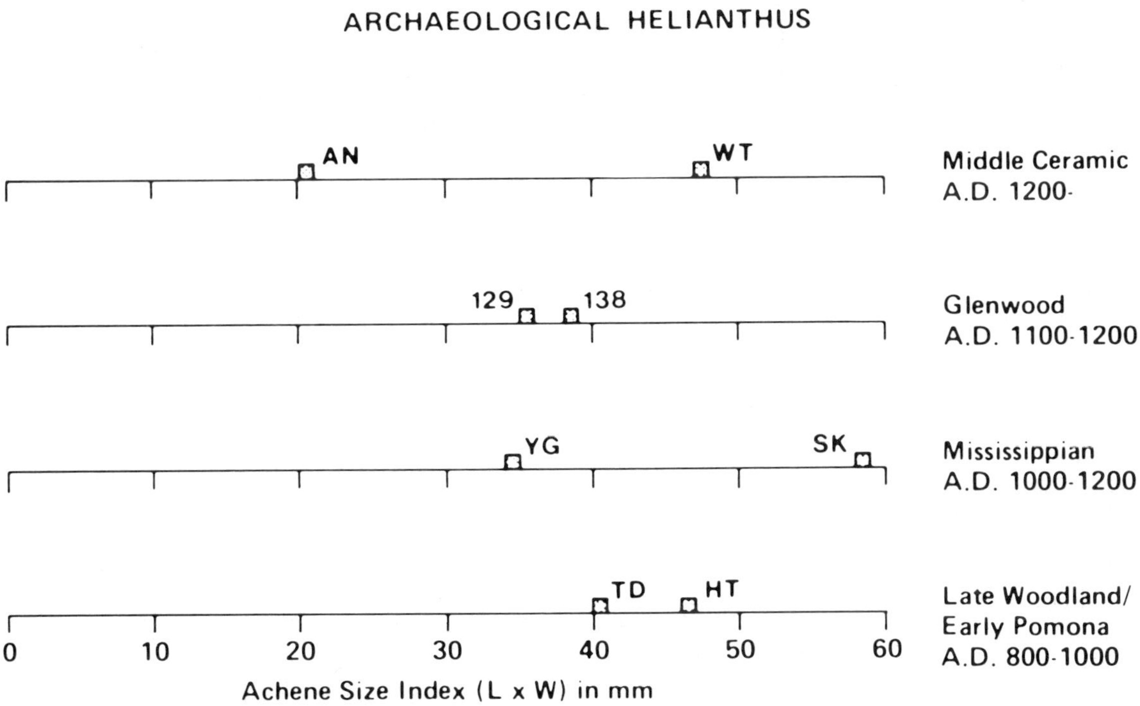

Fig. 3.5 Achene size index (L × W) in mm for archaeological *Helianthus* achenes. Measurements are corrected by Yarnell (personal communication). Site abbreviation key: TD, Two Deer, 14BU55; HT, Hatcher, 14DO19; YG, Young, 23PL4 (Steed-Kisker occupation); SK, Steed-Kisker, 23PL13; 129, 13ML129; 138, 13ML138; AN, Anderson, 14DO32; WT, Witt, 14GE600.

and may remain as a potential food source for several months. Non-cultigen *Iva*, however, disperses its seeds soon after they are mature and they are available for about a two-week period (Asch and Asch 1978). Fifth, *Iva*'s preference for a monocultural environment was not compatible with a multi-crop system. And last, if *Iva* and sunflower were indeed economic competitors, then the relatively greater success in cultivating the latter (Asch and Asch 1982) might account for the ultimate decline of *Iva*.

Zea mays L. (maize)

The development and domestication of maize has received far greater attention in the literature than any other New World cultigen. Several views regarding the evolutionary pathway of this plant species are available (Beadle 1980; Galinat 1985; Mangelsdorf 1983), while additional studies focus on the emergence of specific races of maize and their spread to geographical and ecological niches (Galinat 1985). For Central Plains prehistory it is not necessary to present the opposing views on the biological development of the race *Zea mays*, but rather to discuss the varieties available to Central Plains populations and the time of their arrival.

According to currently available reports, remains of *Zea mays* have been recovered from preceramic contexts in both the Southwest and the Eastern United States, although the exact time maize entered each area remains debatable. Reputed early maize sites in the Southwest are problematic. Maize from Bat Cave, Tularosa Cave, Jemez Cave, Fresnal Shelter, LoDaisKa and other high altitude sites have been reanalyzed according to revised stratigraphic sequences and more reliable dating tech-

Table 3.5. Distribution of *Helianthus* achenes within the Central Plains subarea.

Modern	*N*	*Mean*	*Range*
wild-Douglas Co.	156	4.7 × 2.6 = 12	4.2–5.2 × 2.0–3.1
Sumner Co.	68	4.5 × 2.3 = 11	3.9–5.2 × 1.7–2.7
wild[a]	?		3.0–5.5 × ?
ruderal[a]	?		4.0–7.0 × ?
cultigen[a]	?		6.3–20+ × 3.2–12+
Archaeological site (with map reference number)			
1. Coffey 14PO1[b]	6	4.3 × 2.5 = 11	3.5–4.8 × 2.0–2.8
2. Young 23PL4[b,c]	1	fragment	
3. 14LT304[e]	1	fragment	
4. Hatcher 14DO19[b]	8	10.1 × 4.5 = 46	9.0–11.6 × 3.2–5.4
5. Two Deer 14BU55[b]	120	9.1 × 4.4 = 40	6.5–15.3 × 2.5–7.2
6. Steed-Kisker 23PL13[a]	6	10.8 × 5.3 = 58	9.7–12.7 × 4.3–7.0
7. Young 23PL4[h]	33	8.8 × 3.9 = 34	5.1–12.4 × 2.9–5.4
8. 23CL115[h]	2	6.4 × 3.2 = 21	–
9. 23CL109[h]	1	6.8 × 3.6 = 25	–
10. 13ML139[d,b]	1	11.1 × 3.7 = 41	–
11. 13ML138[d,b]	22	9.8 × 3.9 = 38	6.1–15.6 × 2.5–6.5
12. 13ML129[d,b]	306	9.2 × 3.8 = 35	5.1–14.6. × 2.2–6.1
13. 14BU71[b]	1	9.4 × 3.9 = 37	–
14. 14ML417[b]	3	8.3 × 3.6 = 30	5.6–10.4 × 2.8–4.5
15. 14ML16[f]	1	8.3 × ?	–
16. 14ML15[g]	?	?	–
17. Anderson 14CF508[b]	5	7.0 × 2.9 = 20	6.5–7.7 × 2.6–3.3
18. 14MN509[b]	1	13.0 × 5.4 = 70	–
19. Witt 14GE600[b]	3	9.2 × 5.1 = 47	–

[a] Yarnell 1978; [b] measured by author, corrections made for charring; [c] Kansas City Hopewell occupation; [d] Nickel n.d.; [e] Witty, personal communication; [f] Cutler and Blake 1969; [g] Cutler and Blake 1973b; [h] Blake 1980

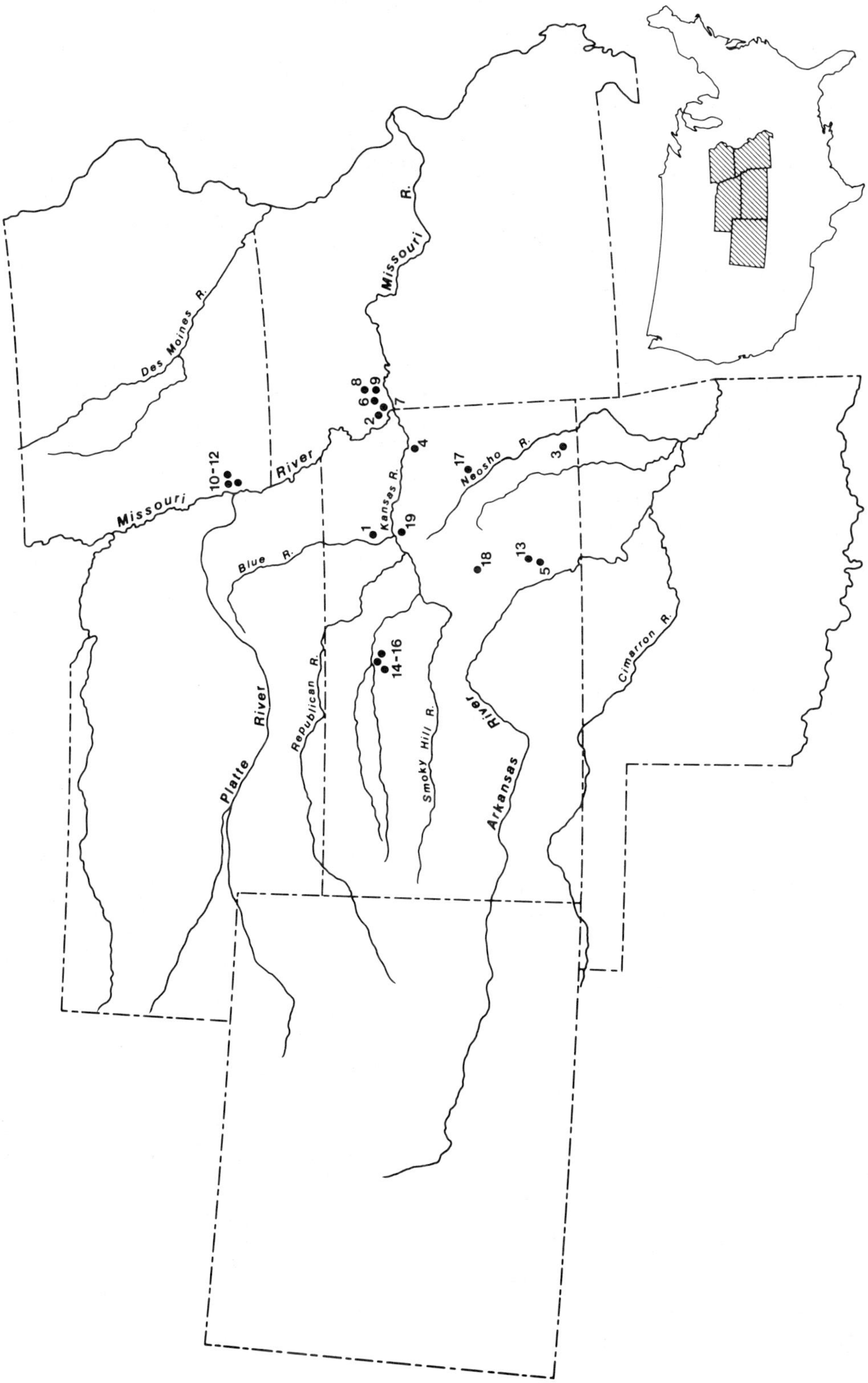

Fig. 3.6 Location of archaeological sites from which *Helianthus annuus* has been recovered. Numbers refer to sites referenced in Table 3.5.

niques to suggest that this cultigen was introduced into the Southwest about 500-750 B.C. (Berry 1985), rather than the earlier suggestion of 2300 B.C. (Mangelsdorf et al. 1967).

All early corn remains from these sites belong to the Chapalote series, identified as a small cob popcorn (Ford 1981; Jones 1949; Mangelsdorf 1974). Varieties of this series have been identified (Cutler and Blake 1973b) although the phenotypes of the early maize in the Southwest reflect considerable genetic diversity (Ford 1981). An additional influence on Southwestern subsistence strategies is recognized with the appearance of a different race of maize about A.D. 200-700. Harinoso de Ocho, Maiz de Ocho, or simply eight-rowed corn, added even greater variability to the gene pool and provided advantages such as higher productivity, easier milling because of the large flour kernels, and an adaptability to more northern latitudes (Galinat 1965; Yarnell 1964). Ford (1981:13) states that this influence provided enough genetic variability to produce all further corn types in the Southwest. This includes the Basketmaker, Pima-Papago and Fremont varieties.

The arrival date of corn in the Eastern Woodlands is apparently open to some debate. Ford (1981) suggests a date of 500 B.C. or roughly equivalent to the accepted date for its arrival in the Southwest. Conard et al. (1984) however, note errors in the relative dating of these remains and suggest an alternate arrival date of 2000-1550 B.P. Asch and Asch (1985) note that the crop was not an important subsistence item in west-central Illinois before A.D. 600, and possibly not even until after A.D. 800. While they note reports of earlier maize in the east dating to about 2000 years ago, they add that the quantities are few and the cultural associations have been questioned.

Reputed early maize from Meadowcroft Rockshelter in Pennsylvania, Daines II Mound in Ohio, and from other early sites in the East has been interpreted to contain some elements of Chapalote flint corn (Blake and Cutler 1982; Ford 1981; Streuver and Vickery 1973). Elements of Maiz de Ocho do not appear until significantly later in time. The significance of the eight-rowed variety is recognized with the emergence of the Northern Flint variety, which came to dominate in the Northeast, Upper Missouri and Mississippi Valleys. It is characterized by a predominance of strongly paired eight rows of large flint kernels. The cobs tend to have a distinct square, pentagonal or hexagonal cross-section while the kernels are usually crescent shaped. In addition the kernels are usually large, but are seldom higher than they are wide. The archaeological samples often lack the starchy endosperm area, accentuating the crescent shape (Yarnell 1964). Some variations are noted, including the presence of 10 and sometimes 12 rows, and soft flint or rarely sweet kernels (Blake and Cutler 1982). The plant is cold and drought resistant and requires a growing season of slightly more than 90 days (Jones 1949). Northern flint varieties, along with lesser amounts of earlier Chapalote-type varieties, became a staple crop initially during the Late Woodland period and most pronounced during the Mississippian period.

The direction from which corn diffused into the Midwest riverine area is largely unknown and still speculative due to lack of good empirical data. The most widely accepted theory suggests it came from the Southwest by way of the Southern or Central Plains (Galinat 1985). According to this theory, the highly genetically diverse corn in the Southwest diffused north and east with natural and cultural selection filtering out certain features from the germplasm which were best adapted to a particular geographical area. With the initial range of variation present in Southwestern maize, there is no need to postulate successive introductions of new corn types (Ford 1981:15). Cutler and Blake (1969) have suggested that the Northern Flint variety could have evolved by a reduction in row number frequencies from 14 and 12 rowed to 8 rowed. This is in agreement with Brown and Anderson (1947:14) who point out that some varieties of popcorn actually resemble the flints in their early maturity, crescent shaped seeds, predominance of eight rows, tillering, and structure of the tassel.

On the other hand, Galinat (1985) states that the 8-rowed maize (Maiz de Ocho) first appeared in the Southwest around A.D. 200-700 from an original location of highland Guatemala or Columbia. This variety was preadapted to high altitudes and northern latitudes and gradually spread eastward along the riverine areas to the Northeast. Certain traits that probably introgressed from the earlier Chapalote race, including the flinty endosperm and the tillering plant habit, helped it adapt to a variety of more northern climates.

The earliest evidence of maize in the Central Plains subarea is found in association with cultivated *C. pepo* and *Iva annua* at the Middle Woodland Trowbridge site. Charred cobs and kernels were recovered from a refuse pit and were dated by seriation of associated ceramics (Johnson and Johnson 1975) to approximately A.D. 250-400 (Bell 1976). A small sample was described by Cutler and Blake as a "moderately thickened cob, less than northern flints, but not an open early type cupule" (1973b:32). The isolated kernels are from mainly 10-rowed ears although at least some are from 12-rowed varieties. A second larger sample was recently analyzed by Richard Ford at the University of Michigan who provided additional information on the cultigen. The maize belongs to the Chapalote series, although it does not have an accepted varietal name in the Central Plains. The kernels are small, triangular shaped and thin in cross-section (Table A.4). According to Ford, this maize is the earliest series known from the Central Plains and was not as productive or as adapted to the local environment as the later varieties.

Corn recovered from the Renner site, also a Kansas City Hopewell site (Roedl and Howard 1957; Wedel 1943), is presumed to be of the same variety although only two kernels were recovered. Additional evidence of corn cultivation in a possible Middle Woodland context comes from the southeastern section of Kansas. Site 14LT304, a Cuesta phase occupation, yielded two maize kernels. Given the postulated association between the Kansas City area Middle Woodland populations and the Cuesta phase populations, it is possible that the same variety of corn was being grown by both. The Chapalote series popcorn continued to be grown in small quantities until historic times. It appears at the Lawson site, a Loeske Creek Woodland occupation in northern Nebraska (Fig. 3.7) as well as in several Middle Missouri sites (Cutler and Blake 1973b).

Following the decline of the Kansas City Hopewell until about A.D. 1000, the evidence of maize cultivation in the Central Plains is limited. Four sites offer data on maize cultivation during the latter part of the Early Ceramic period. A single maize kernel fragment was recovered from the Stiles site (14MM13) in eastern Kansas. Assigned to the Plains Woodland Greenwood phase (Blakeslee and Rohn 1982), the occupation is dated to A.D. 725 (Table 2.2). One maize kernel was also identified from site 14BU57 in southcentral Kansas (M. Brown 1982b). However, the association of this cultigen in an Early Ceramic context is difficult to ascertain. The corn was recovered from a feature also yielding charcoal dated to A.D. 1195±235. The date may reflect a later, but extremely limited, occupation at the site since most of the ceramics and projectile points recovered are styles similar to those from earlier Butler phase Woodland sites in the area. Additionally, the large sigma makes the date somewhat suspicious. If additional finds of maize in Plains Woodland contexts are made, the temporal association of the botanical data from this site may be less controversial.

Excavations at Hatcher (14DO19) in northeastern Kansas provide additional information on maize cultivation. Two kernels were identified, along with squash and sunflower, from the contents of a single feature. The Pomona occupation reflected in the artifacts is dated to A.D. 875 (Table 2.2). However, based on a calibration of the date and an analysis of the ceramics and lithic styles, Brown (1984) believes the occupation to be about 100 years later than the date indicates.

A final, and also speculative, association of corn in a Woodland context comes from sites in the Smithville reservoir area of northwestern Missouri (McHugh 1982). Several maize kernels were identified from sites 23CL274 and 23CL276. These multi-component Late Woodland/ Steed-Kisker sites also yielded cultivated *Iva* and, as argued earlier the temporal association of these cultigens cannot be determined.

At present, a postulated use of maize in late Early Ceramic occupations seems somewhat conjectural based on the botanical evidence (Table 3.6). Such a use is not totally unlikely however, and future excavations with proper recovery techniques for botanical data will hopefully resolve this problem. The importance of finding maize in a good Woodland context focuses primarily on the variety cultivated at that time. Although the sample size of maize kernels from the above sites is small, they exhibit characteristics different from the earlier Chapalote series corn. The kernels are crescent shaped, lack the starchy embryonic section, and are somewhat short and squat. This suggests that by the end of the Woodland period, approximately A.D. 800-1000, an eight-rowed influence had either developed or was introduced in the Central Plains. This influence may have been critical in providing an environmentally stable crop in the area.

The suggestion that this influence occurred prior to the Middle Ceramic period is logically supported by the fact that several groups throughout the study area document the presence of a new variety in their habitation refuse by A.D. 1000 or even slightly earlier. Data from the Solomon River, Bemis Creek, and Steed-Kisker complexes demonstrate the presence and cultivation of an 8-14 rowed variety of maize whose characteristics include some of those assigned to the Northern Flint race. Additionally, the importance of maize cultivation at these complexes is suggested by the quantity of remains recovered (Table 3.6) and the presence of other cultigens.

Maize fragments from five Solomon River phase sites have been described as mainly medium to small kernels from 8, 10, and 12 rowed varieties (Cutler and Blake 1969, 1973b:31-32). Corn samples for two other sites, 14ML417 and 14JW301, have not been analyzed. The sites of this complex have radiocarbon dates of A.D. 830-A.D. 1190 (Table 2.2). Charred corn from the Bemis

Table 3.6. Distribution of *Zea mays* in Central Plains subarea.

Archaeological Site with Map Reference Number	*Sample Size**	*Cultural Affiliation*
1. 14WY1 Trowbridge	100+ kernels, 1 cob	Kansas City Hopewell[a]
2. 23PL1 Renner	2 kernels	Kansas City Hopewell[b]
3. 14LT304	2 kernels	Cuesta[c]
4. 14DO19 Hatcher	2 kernels	Pomona[d]
5. 14MM13 Stiles	1 kernel	Greenwood[e]
6. 14BU57	1 kernel	Butler[f]
7. 14BU55 Two Deer	114 kernels	Bemis Creek[a]
8. 14HP5 Anderson	13 cobs	Bluff Creek[c]
9. 14SR303 Buresh	63 kernels, 33 cobs	Bluff Creek[c]
10. 14HP1 Anthony	600+ kernels, 2 cobs	Bluff Creek[g]
11. 14SR305 Nulik	?	Bluff Creek[c]
12. 23CL? Cloverdale	?	Steed-Kisker[t]
13. 23PL6 Vandiver	kernels	Steed-Kisker[g]
14. 23PL54 McClaron	kernel and cob	Steed-Kisker[g]
15. 23CL226	3 kernels	Steed-Kisker[r]
16. 23CL274	3 kernels	Steed-Kisker[r]
17. 23CL276	4 kernels	Steed-Kisker[r]
18. 23CL113 Friend and Foe	32 kernels, 8 cobs	Steed-Kisker[h]
19. 23PL13 Steed-Kisker	kernels and cobs	Steed-Kisker[g]
20. 23PL4 Young	19 kernels, 125+ cob	Steed-Kisker[h]
21. 23PL48 Gresham	kernels and cob	Steed-Kisker[g]
22. 23CL118 Butcher	?	Steed-Kisker[s]
23. 23CL115	13 kernels, 20 cobs	Steed-Kisker[h]
24. 23CL109 Hulse	47 kernels, 74 cobs	Steed-Kisker[h]
25. 23PL16 Coons	235+ kernels, 14 cobs	Steed-Kisker[h]
26. 23PL80 White	105+ cobs	Steed-Kisker[h]
27. 23JA115 Seven Acres	2 kernels	May Brook[i]
28. 23JA38 Bowlin Bridge	1 kernel	May Brook[j]
29. 14CF508 Anderson	5 kernels	Pomona[k]
30. 14AT2	1 kernel	Pomona[u]
31. 13ML126	14 kernels, 1 cob	Glenwood[l]
32. 13ML128	12 kernels, 2 cobs	Glenwood[l]
33. 13ML129	85 kernels, 20 cobs	Glenwood[l]
34. 13ML130	17 kernels, 1 cob	Glenwood[l]
35. 13ML135	20 cobs	Glenwood[l]
36. 13ML136	8 kernels	Glenwood[l]
37. 13ML137	13 kernels, 10 cobs	Glenwood[l]
38. 13ML138	200 kernels, 24 cobs	Glenwood[l]
39. 13ML139	90 kernels, 3 cobs	Glenwood[l]
40. 14DP13 Squaw Creek	cob fragments	Nebraska[g]
41. 140B27 Sumter	150 kernels	Solomon River[g]
42. 14ML417	kernels and cobs	Solomon River[c]
43. 14ML11	3 kernels	Solomon River[g]
44. 14ML15 Thull	160 kernels, 2 cobs	Solomon River[g]
45. 14ML16	few kernels	Solomon River[g]
46. 14ML371 Long	few kernels	Solomon River[g]
47. 14JW301 Bud Hill	?	Solomon River[c]
48. 14CY102	100+ kernels, 1 cob	Upper Republican[g]
49. 14NT301	?	Upper Republican[c]
50. 25FT35 Mowry Bluff	kernels and cobs	Upper Republican[n]
51. 14LC301 Root	?	Smoky Hill[c]
52. 14OT5 Minneapolis	kernels and cobs	Smoky Hill[p]
53. 14RY401 Griffing	1 cob fragment	Smoky Hill[h]
54. 14GE600 Witt	30 cobs	Smoky Hill[v]
55. 14PO4 Budenbender	1 kernel	Smoky Hill[g]
56. 14MH42 Walls	kernels and cob	Smoky Hill[g]
57. 14CY1 Mugler	463 kernels, 1 cob	Smoky Hill[m]
58. 14CY4 Kemp	1182 kernels, 7 cobs	Smoky Hill[m]
59. 14CY30 Woods	kernels and cobs	Smoky Hill[o]

Table 3.6. Distribution of *Zea mays* in Central Plains subarea. (continued)

Archaeological Site with Map Reference Number	*Sample Size**	*Cultural Affiliation*
60. 14CY? Ware	209 kernels	Smoky Hill[m]
61. 14GE21	kernels and cobs	Smoky Hill[g]
62. 14GE127	few kernels	Smoky Hill[g]
63. 14PT1 Pratt	kernels and cobs	Pratt[g]
64. 14RH301	?	Middle Ceramic[c]
65. 14MH504	20 kernels, 28 cobs	Great Bend[m]
66. 14MH509	3 kernels, 2 cobs	Great Bend[m]
67. 14MH328	28 kernels, 6 cobs	Great Bend[m]
68. 14RC8 Tobias	kernels and cobs	Great Bend[g]
69. 14RC5 Malone	kernels and cobs	Great Bend[g]
70. 14RC306 Hayes 4	kernels and cobs	Great Bend[c]
71. 25PT12 Lawson	6+ kernels	Loeske Creek[q]

[a] analyzed by Richard Ford, described in this study; [b] Roedl and Howard 1957; [c] Witty, personal communication and this study; [d] Johnson 1968; [e] Blakeslee and Rohn 1982; [f] Brown 1982b, and described in this study; [g] Cutler and Blake 1973b; [h] Blake 1974, 1980; [i] Brown 1981; [j] Schmits 1982; [k] Rohn et al. 1977 and described in this study; [l] Nickel n.d.; [m] Adair, this study; [n] Cutler and Blake 1969; [o] Witty 1963; [p] Wedel 1935; [q] Kivett 1952; [r] McHugh 1982; [s] Calabrese 1969; [t] Shippee 1972; [u] Williams 1986; [v] Adair 1984b

* cob designation refers to cob fragments and cupules.

Creek occupation at Two Deer also represents a mixture of 8-12+ rows that may have been evolving towards 8-row dominance (Richard Ford, personal communication). This occupation dates to A.D. 860-A.D. 1060 (Table 2.2). The Steed-Kisker corn is from 8, 10 and some 12 row varieties which Cutler and Blake (1973b:42-43) label as Northern Flint, "but not an extreme form." Elsewhere, they discuss the possibility that the 8-row Northern Flint variety mixed with other, and probably older, varieties in the Central Plains to produce a cob with a higher average row number (Blake and Cutler 1982). Maize samples not included in Cutler and Blake's analysis show 8-14 rowed cobs with cupule widths of 5-10 mm (Adair 1984b).

The variability observed in the corn from these complexes continues throughout the Middle Ceramic period. Eight to 14 rowed corn, with cupule widths of 4-10 mm, has been identified in the Upper Republican sites 25FT35, 14ML417, 14ML15, and 14GY102 (Cutler and Blake 1969, Adair 1984b). Nebraska phase sites, including the Glenwood locality, have also yielded 8-14 rowed maize that exhibit measurable cupule widths of 5-10 mm (Nickel n.d., Adair 1984b). Corn recovered from the Bluff Creek occupations recorded at sites 14HP5 and 14SR303 is primarily an 8-12 rowed variety with a range in cupule width of 4-11.5 mm. Smoky Hill phase sites exhibit a similar pattern. Corn recovered from 14GE600, 14CY1, 14CY4, and 14GE21 belongs to a 8 to 14 rowed variety with cupule widths ranging from 6.5 to 12 mm (Adair 1984b).

The possibility that an earlier variety, perhaps a Chapalote series, was also grown during this time is indicated by "triangular, longer than wide, medium size . . . thin grains from 14RC8", the Tobias site (Cutler and Blake 1973b:30); 14-rowed cobs from Mowry Bluff (25FT35) and 14CY102 (Cutler and Blake 1969) as well as some Glenwood sites (Nickel n.d.); small thin triangular kernels from Smoky Hill site 14CY4 and the Bluff Creek site 14HP5 (Table A.5); and a few popcorn-like kernels from 23PL48, a Steed-Kisker occupation (Cutler and Blake 1973b). The same

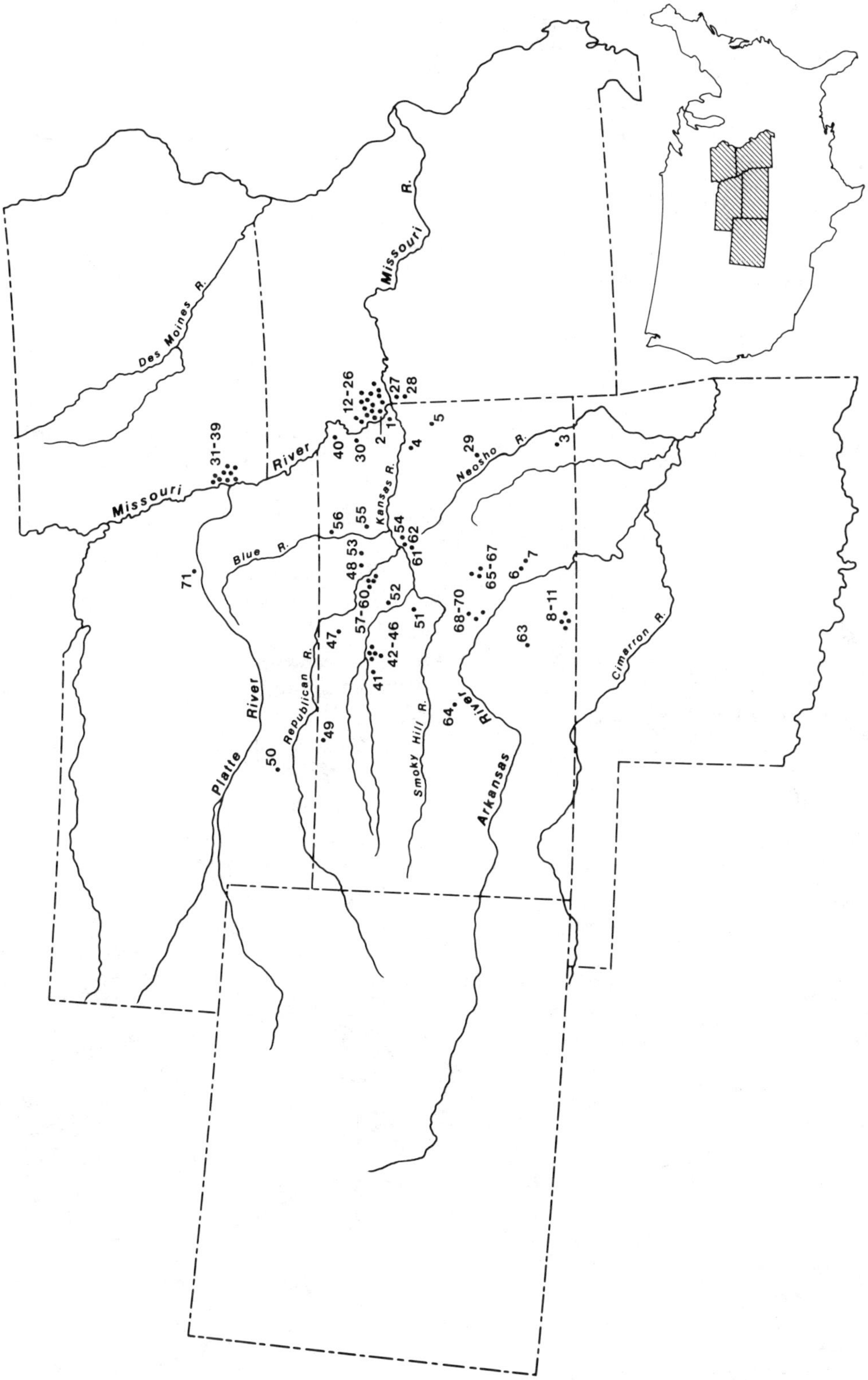

Fig. 3.7 Location of archaeological sites from which *Zea mays* has been recovered. Numbers refer to sites referenced in Table 3.6.

variability suggestive of different varieties is also documented from Cahokia to the east of the study area (Cutler and Blake 1973a). The fact that different types or varieties are present in the Central Plains and surrounding areas even in historic periods argues against the earlier Middle Ceramic remains being intermediate between some early variety and the Northern Flint. Rather, the data indicate that several varieties either developed in the Central Plains as a result of environmental and human selection or were introduced from elsewhere and were grown simultaneously by prehistoric inhabitants. If there was a tendency by these peoples to select for a variety dominated by eight rows and similar to the Northern Flint, the present data are unsupportive.

Phaseolus vulgaris L. (Common bean)

Although about 80 species of the genus *Phaseolus* are native to the New World, only four have been prominent as food crops. Of these four, the common bean, *P. vulgaris*, was the only one cultivated in the Central Plains and Eastern Woodlands (Kaplan 1971). This species includes such varieties as Navy, Red Kidney, and Pinto, and both bush and vine forms of the plant. The distinctions between green, string, or snap beans and dry or field beans is based on use rather than biological characters (Kaplan 1971: 517).

Beans have long been noted for their nutritional significance as a source of vegetable protein. Additionally, most of the varieties and species of *Phaseolus* are high in the amino acids lysine and tryptophane and it is precisely this factor that results in the nutritional complement that exists between most bean and corn varieties. Although high in total protein content, maize has been demonstrated to be deficient in both of these amino acids (Kaplan 1973). Therefore, in a diet where maize is the main foodcrop, the addition of beans counteracts the maize deficiencies and results in the achievement of a protein intake of high biological value. The close relationship between beans and corn has been found

Table 3.7. Distribution of *Phaseolus vulgaris* in Central Plains subarea and surrounding regions.

Archaeological Site with map reference number	*N*	*Mean L × W × T*	*Cultural Affiliation*
1. 23PL1 Renner[a]	1	modern pinto	Kansas City Hopewell
2. 23HI172 Blackwell Cave[b]	?		Village period
3. 14CF506 Winn or 14CF508 Anderson[c]	?		Pomona
4. 14RC5 Malone[e]	?		Great Bend
5. 14NT301[e]	?		Upper Republican
6. 14DP13 Squaw Creek[f]	200	11.0 × 6.2	Nebraska
7. 14ML417[e]	1	fragment	Upper Republican
8. 23CL115[g]	15	8.9 × 5.1 × 3.7	Steed-Kisker
9. 23CL109[g]	2	8.5 × 4.8	Steed-Kisker
10. 23PL4 Young[g]	2	fragments	Steed-Kisker
11. 23PL16[g]	74	8.7 × 5.0 × 4.1	Steed-Kisker
12. 14MN504[d]	1	fragment	Great Bend
13. 13ML126[h]	2	fragments	Glenwood
14. 13ML129[h,d]	11	9.5 × 5.4 × 3.5	Glenwood
15. 13ML138[h,d]	65	10.4 × 6.8 × 5.2	Glenwood
16. 13ML139[h]	7	fragments	Glenwood

[a] Wedel 1943; [b] Falk 1969a; [c] Rohn et al 1977; [d] Adair, this study; [e] Witty, personal communication; [f] Heavin 1970; [g] Blake 1974, 1980; [h] Nickel n.d.

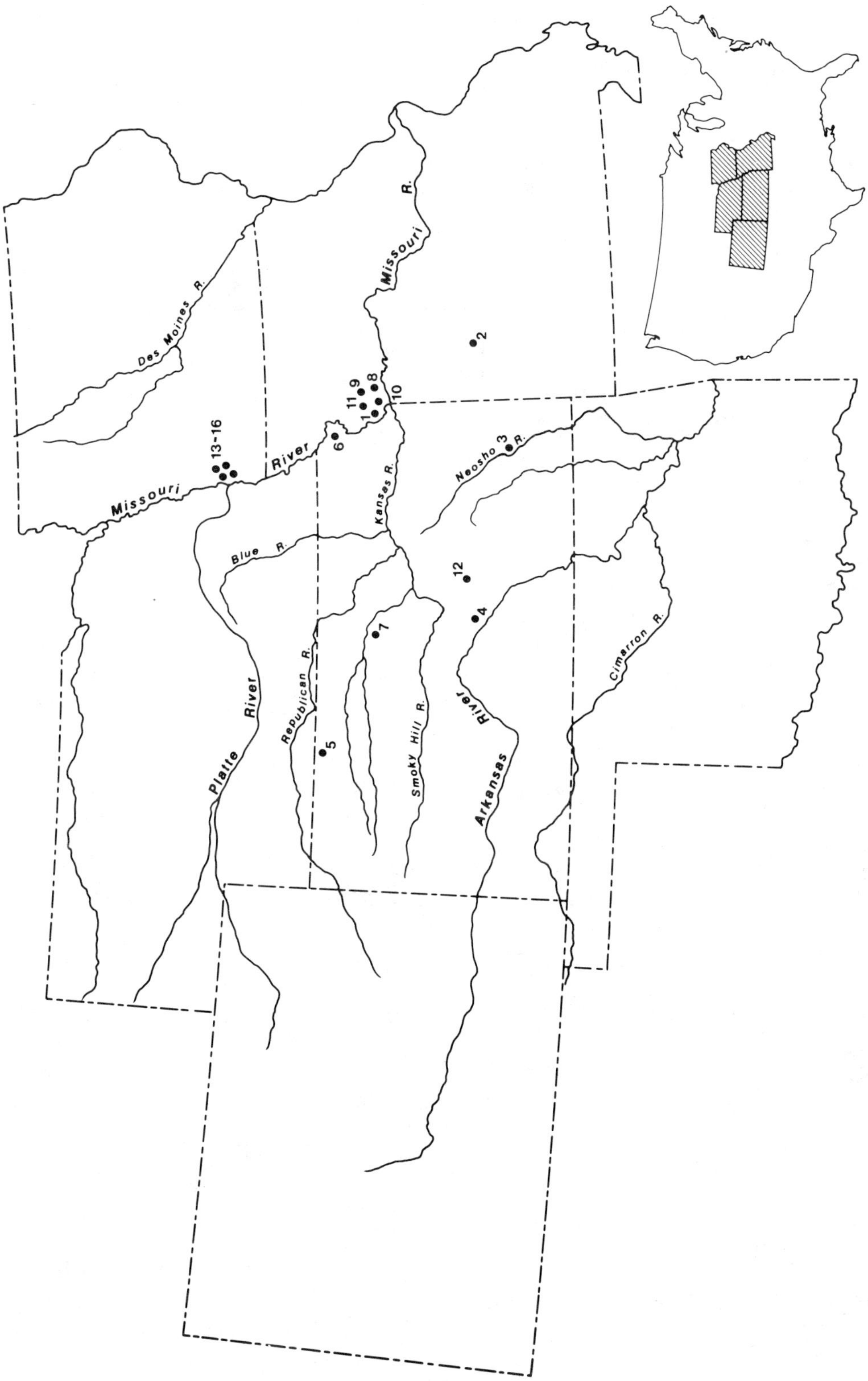

Fig. 3.8 Location of archaeological sites from which *Phaseolus vulgaris* has been recovered. Numbers refer to sites referenced in Table 3.7.

in many pre-Columbian populations throughout Mesoamerica, the Andean region, and North America. Kaplan (1971, 1973) feels that this combination came about as a result of the beans yield, seasonality of harvest, and overall compatibility with other crops as well as the nutritional agreement.

The earliest appearance of the common bean north of Mexico is in Southwestern sites. It appears after corn, squash, and gourd but the date is uncertain. Early reports indicate its association with maize at about 2300 B.C., although a re-examination of the dating and stratigraphic columns at these sites suggests a more conservative arrival time of 300-500 B.C. (Ford 1981:14). This cultigen is found initially in the Mogollon area from where it dispersed to other areas in the Southwest, northward into the Durango area, and eventually eastward after A.D. 800 (Yarnell 1976; Ford 1981). The route followed in the eastern dispersal is still undefined.

The earliest, but controversial, evidence of *P. vulgaris* in the Central Plains is found in a Middle Woodland Hopewell context at the Renner site (Wedel 1943). Beans resembling the modern pinto were recovered during excavations but rapidly disintegrated and were never examined by a botanist. The identity of these remains as beans continues to be disputed (Ford 1979) and it is believed that they may actually have been honey locust seeds.

The arrival of the common bean is documented from the American Bottoms sometime after A.D. 950 (Cutler and Blake 1973b; Johannessen 1981; Asch and Asch 1982) and following the adoption of maize agriculture, became a significant subsistence item. Data from the Central Plains would suggest a similar development but occurring a bit later (Table 3.7). Common beans were recovered from several Steed-Kisker sites, such as 23CL115, 23CL109, 23PL16 and Young (23PL4) with acceptable dates ranging from A.D. 1070 to A.D. 1315 (Table 2.2). Slightly larger beans (Table 3.7) were identified from the Squaw Creek site (Heavin 1970), a Nebraska occupation. Four Glenwood sites (Fig. 3.8) also include the common bean in their archaeobotanical record. Beans are reported from two Upper Republican sites, 14NT301 and 14ML417 as well as two Great Bend occupations, Malone (14RC5) and 14MN504 (Table 3.7). The small sample sizes of this cultigen, along with its relatively limited geographical distribution (Fig. 3.8), is most probably due to the rarity of it being burned and/or its limited utilization at the time.

Chronological Summary

A summary of the evidence for cultigens in the Central Plains subarea is presented in Figure 3.9. The earliest secure evidence of cultivation in the study area dates to the Middle Woodland Hopewell occupation. Squash, corn, and *Iva* were cultivated by perhaps A.D. 250. There is no evidence as yet to suggest that any of these plants were important to previous Archaic populations or that, in the case of *Iva*, domestication was the result of a long process or use. The common bean is known to be a cultigen of eleventh century and later populations. The native sunflower reached domestication status at about A.D. 900 and apparently replaced the use of *Iva* as it became a dominant cultigen to Middle Ceramic populations. The bottle gourd is not known in an archaeological context until about A.D. 1000 although it is documented from sites surrounding the Central Plains dating several centuries earlier. Although corn was grown by Middle Woodland peoples, a more environmentally adapted variety of this cultigen appeared or developed on the Plains around A.D. 800-1000, resulting in an increased use of this plant as a major food item. At the time of European contact, many populations throughout the Central Plains manifested a subsistence of the maize-beans-squash-sunflower assemblage.

The temporal summary illustrated in Fig. 3.9 suggests a hiatus period in which at least three domesticated plants are absent from the archaeological record for a period of 150 to 550 years following their first appearance as cultigens. Squash was recovered from only one Kansas City Hopewell site and does not appear again in botanical samples until A.D. 900. The early Chapalote variety of corn grown by the Kansas City Hopewell is present in only a few sites and appears to be a subordinant variety in later period sites. *Iva* was apparently a more important subsistence item to Kansas City Hopewell populations, as it is found in all three Middle Woodland phases. Our current knowledge of the situation suggests that *Iva* may have been the only Middle Woodland cultigen to have experienced an uninterrupted sequence of domestication from the Middle Woodland to the Mississippian period, or from about A.D. 250-1200. However, this is a tentative suggestion given the questionable association of *Iva* with Late Woodland occupations, hence the gap in the summary of *Iva*.

The hiatus period in agricultural activity could be attributed to several factors, which are not necessarily mutually exclusive. First, there is a spatial and cultural separation in the evidence for agriculture. Prior to A.D. 900, almost all cultigens are associated with the Middle Woodland period primarily in the Kansas City area, while after this time, cultigens are found throughout the study area. Middle Woodland Kansas City Hopewell populations may not have sufficiently dispersed cultigens throughout the study area during the earlier Trowbridge and Kansas City phases since these foods were only redundant subsistence items and not staples in the economy. Consequently, Plains Woodland populations, which temporally overlapped terminal Kansas City Hopewell, received domesticated plants from some other source at a much later date

Second, poor preservation, inadequate recovery, and insufficient identification of archaeobotanical remains have severely limited our understanding of the availability and importance of cultigens at particular times. These situations may be more pronounced in Plains Woodland sites since these are frequently located in plowed fields and are often thin deposits.

A third explanation identifies two temporally and spatially separate periods of agriculture adoption in the Central Plains. The earliest period corresponds with the Kansas City Hopewell florescence when the tropical cultigens of corn, squash, and the native cultigen *Iva* were grown as casual subsistence items. There is no evidence to suggest these foods even increased in importance during the latter Kansas City Hopewell phases from their initial introduction. Use of domesticated plants declined considerably (per-

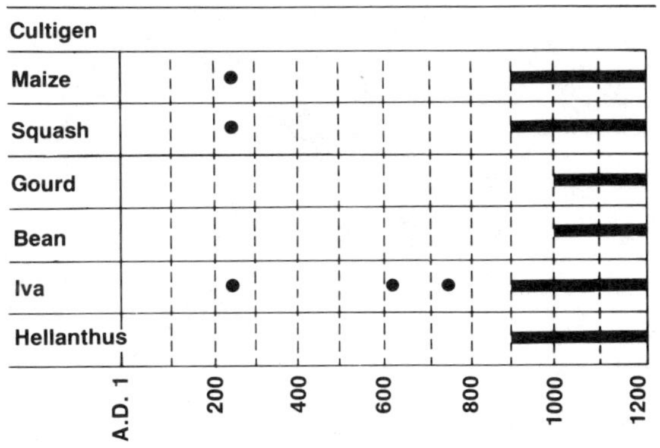

Fig. 3.9 Chronological summary of cultigens in the Central Plains subarea.

haps totally) following the Kansas City Hopewell period until about A.D. 800-900. The second period of cultivation use may indeed mark the "horticultural revolution" (cf Asch and Asch 1982) of the Central Plains. Beginning about A.D. 800-900, a new variety of maize appears on the Plains and by A.D. 1000, it is recorded in sizeable quantities from archaeological sites along with squash, sunflower, and marshelder. Shortly after, the picture was complete with the arrival of the common bean.

If future excavations are successful in recovering cultigens that eliminate the temporal gap, then a fourth factor can be suggested. Ignoring the spatial dimension, the presence of cultigens can be assigned to two levels of importance within their respective subsistence strategies. An initial acceptance of cultigens is contingent upon how well these foods fit into pre-existing categories of subsistence items. Cultigens which are no more important to a group than several wild resources are part of a domesticatory, or incipient agricultural, system, while cultigens which are staples and therefore intensively planted and harvested constitute an agricultural system (Rindos 1980). Quantified botanical data as well as related cultural factors (such as technological, demographic, social) are often needed to distinguish between the two systems. In the following chapters, a model for the development of agriculture in the study area is proposed and the explanations proposed to account for the agricultural hiatus period are discussed.

CHAPTER 4

The Development and Importance of Agricultural Systems

A theoretical model for the development of agriculture, presented in Chapter 1 of this study, stressed certain factors as importantly influencing changes in prehistoric subsistence. The validity of each of these factors as significant contributions in the overall change in subsistence strategies in the Central Plains would require extensive research well beyond that intended for this study. The objective here is to put forth information crucial to this research with the realization that this empirical data is not intended to be a well developed model. The botanical record has provided the basis for this chapter which discusses the varying levels of importance of food production in Central Plains subsistence strategies. Keeping in mind the size of the study area however and the relatively little amount of work conducted in some regions, the empirical data may seem to be insufficient as a support to the theoretical model. Ironically, this is a beneficial consequence for it demonstrates the need to generate testable hypotheses from specific data sets. This is particularly true for Central Plains prehistory where the work toward understanding major developments, such as the beginnings of agriculture, is very much in its infancy. The botanical record alone emphasizes this and while future excavations along with better and more consistent recovery techniques will certainly enhance the empirical data base, their context within a developed hypothesis will synergistically refine our understanding.

Early Ceramic Domesticatory Systems:

Middle Woodland

At present, the earliest evidence of domesticated plants in the Central Plains is in post-Archaic contexts. If (and more likely when) future excavations are successful in identifying cultigens in Archaic sites, they along with the Early Ceramic Middle Woodland sample will represent food items casually added into a basic hunting and gathering mode. They contributed added security against failure of native nut masts and seeds and did not become a staple because there was no element of advantage. The subsistence strategies of the Kansas City Hopewell included a reliance on large ungulates, especially deer, gathering of nut harvests and native flora, which were dependable and reliable resources, and a limited cultivation of corn, squash, and *Iva* (Adair 1977, this study; E. Johnson 1975; Robinson 1976). This strategy supported people in the area for 600-800 years without the need for increasing the availability of the domesticates. Several lines of evidence are presented as supportive of this suggestion. These may also strengthen the argument that similar subsistence strategies existed between the Kansas City and Illinois Valley Hopewell populations.

As discussed in Chapter 2, the Hopewell populations in the lower Illinois River Valley are believed to have exerted an as yet undetermined amount of influence on the emerging Middle Woodland cultures in the Kansas City region. The selection of the latter area for occupation may have been influenced in part by the availability of a suitable raw material, chert, for the manufacture of diagnostic Middle Woodland artifacts (Kay 1979; Reid 1976) and the rich and abundant resource base provided by the ecotonal setting. Such an ecological parameter provided microenvironments suitable for resource exploitation that were similar to those of the Illinois valley (Reid 1976). Such similarities helped perpetuate a well established Hopewell tradition.

Documented subsistence strategies of Middle Woodland populations in Illinois (Asch et al. 1979), in the Big Bend region of central Missouri (Robinson 1976; K. Brown 1975a; Kay 1975) and in the Kansas City area (Adair 1977; K. Brown 1975b; E. Johnson 1975; O'Brien 1982) are very similar (cf. Ford 1979: Table 29.1). Not only are the floral and faunal remains indicative of a hunting and gathering economy supplemented with cultigens, but the species of plant and animals exploited by each regional group are remarkably similar. As a way of demonstrating this similarity in native floral exploitation, an Index of Similarity was computed using three Kansas City Hopewell sites (Trowbridge, Young, Yeo), one central Missouri site (Fisher-Gabert), and five Illinois Valley occupations (Dickson Camp, Pond, Newbridge, Carlin, and Weitzer). This index, used commonly in ecological studies (King 1982b), is derived by the formula:

$$\text{Index of Similarity } \frac{2w}{a+b}$$

Where: "a" is the number of taxa in one sample.

"b" is the number of taxa in another sample.

"w" is the number of taxa both samples have in common.

The range in similarities is from 0 (entirely different) to 1 (identical). The results of this study (Table 4.1) suggest a varying amount of similarity between the taxa of flora utilized by the various regional Middle Woodland populations, with scores ranging from .13 to .71. Interestingly, the lowest scores are those resulting from the comparison of the Kansas City and central Missouri sites, a situation best attributed to the varying attention paid to flotation and floral identification at these sites. For example, two low scores of .13 and .17 involve the taxa from Fisher-Gabert which do not include the presence of any nut masts, a resource quite abundant in the Young and Trowbridge samples. A possible explanation for the lack of these remains at Fisher-Gabert may be that only light fractions of flotation matrices were analyzed, since no floral remains of any size or weight are reported by Robinson (1976). It seems only logical to assume that various nuts were available to prehistoric populations occupying the floodplain forest setting of the Fisher-Gabert site; the presence of walnut (*Juglans nigra*) and hickory (*Carya spp.*) shells were noted in a cursory examination of unsorted matrix samples from the Mellor site, a Middle Woodland occupation located only 25 miles from Fisher-Gabert.

Low scores are also noted from comparisons of Illinois Valley and Kansas City Hopewell, although they are quite infrequent. Keeping in mind the temporal range and spatial distribution of sites reflected in the nine occupations compared, a few low scores are to be expected. The majority of scores however, reflect a strong similarity in the choice of floral resources selected by the different regional groups, a similarity perhaps explained by an underlying common exploitation pattern between the areas.

If populations were indeed migrating west out of the Illinois River Valley, they chose to retain their developed and efficient subsistence strategies by selecting geographic regions which could support such a pattern. While some regional variations in the distribution and abundance of resource items were responsible for differences between regions, there was an attempt to keep a similar food procurement system. This has also been noted in the faunal remains. The white tailed deer is the most common animal represented in the various Hopewell sites, regardless of their location. There was apparently little Middle Woodland utilization of the bison by Hopewell groups on the Plains border.

Similarities noted in the sizes of *Iva* achenes between the Kansas City and Illinois Valley areas (See Table 3.4) provides additional evidence of a common subsistence strategy. As discussed in the previous chapter, these similarities may be reflective of a parallel involvement in the domestication of this species. At present, there is little evidence to indicate that factors promoting this involvement developed out of preceding Archaic or Early Woodland populations in the Kansas City area; the suggestion has already been made that the intensity of such horticulture was part of the subsistence strategy of populations to the east which was acquired by groups in the Kansas City area. Whether this was due to contact or direct migration cannot be answered at this point.

Table 4.1. Index of Similarity between Floral Remains from Kansas City, Central Missouri, and Illinois Valley Hopewell Sites.

	Trowbridge[a]	*Young*[b]	*Fisher-Gabert*[c]	*Yeo*[d]	*Dickson Camp*[e]	*Pond*[e]	*Newbridge*[f]	*Carlin*[f]	*Weitzer*[f]
Trowbridge	–	.21	.13	.26	.34	.63	.47	.32	.43
Young		–	.17	.50	.43	.36	.54	.46	.43
Fisher-Gabert			–	.32	.28	.16	.45	.36	.34
Yeo				–	.66	.71	.37	.42	.66
Dickson Camp					–	.55	.43	.48	.63
Pond						–	.34	.36	.45
Newbridge							–	.32	.26
Carlin								–	.55
Weitzer									–

[a] Johnson 1972; [b] Adair 1977; [c] Robinson 1976; [d] King 1982a; [e] Asch and Asch 1980; [f] Asch and Asch 1981

Presenting data indicative of similarities in settlement patterns (A. Johnson 1974, 1983), lithic styles (Heffner 1974; Montet-White 1968; Reid 1976), ceramic designs (A. Johnson 1979; Kay and Johnson 1977), and subsistence practices between the Kansas City, Missouri, and Illinois Valley Hopewell provides a basis from which the importance of domestication can be assessed. It has already been implied that domesticated plants maintained a relatively insignificant importance level throughout the Kansas City Hopewell duration. No botanical data exist to demonstrate that any of the cultigens increased in importance and *Iva* is the only cultigen currently known to be present in all three Kansas City Hopewell phases. Extensive flotation and vegetal analyses in the Illinois River Valley (Asch and Asch 1980, 1981) and in the American Bottoms (Johannessen and Whalley 1980, Johannessen 1981) provide supportive data suggesting the importance of native flora over cultigens in the Middle Woodland period. However, these studies also demonstrate the importance of four starchy taxa; *Chenopodium berlandieri*, *Phalaris caroliniana*, *Hordeum pusillum*, and *Polygonum erectum*, which Asch and Asch (1980, 1982, 1985) convincingly argue were cultivated. Maygrass (*Phalaris caroliniana*) and barley (*Hordeum pusillum*) have not yet been identified from a Kansas City Hopewell site and while both knotweed (*Polygonum spp.*) and goosefoot (*Chenopodium spp.*) are present in samples, the species level of these remains have not been determined. Forthcoming studies may demonstrate the presence of the domesticated species of these starchy annuals in Kansas City Hopewell sites, a conclusion that could importantly influence our understanding of the significance of agriculture during this period. As previously discussed however, botanical data alone may be insufficient evidence on which to base major conclusions on subsistence, especially if flotation techniques have not been consistent among sites.

Another approach to the question of the importance of domesticated plants in the Middle Woodland period would be to demonstrate that animal rather than plant foods were the primary source of protein and calories and that a hunting-gathering strategy prevailed, with modification, when population densities threatened the balance between people and resources.

It has been suggested that the incorporation of a rich and reliable aquatic resource base into a seasonal procurement strategy would be a sufficient condition to permit the establishment of a sedentary lifestyle (Reid 1980b). The Kansas City Hopewell populations however apparently managed to acquire and retain a certain level of sedentism without relying on either a rich or a reliable aquatic resource base. Poor preservation of fish remains may account for some of the low representation, although Reid has demonstrated that the ". . . marshes and backwaters of the Missouri River Floodplain did not attract, impound, or produce aquatic resources . . . in concentrations sufficient to sustain a seasonally or permanently nucleated population of foragers" (Reid 1980b:33). Likewise, migratory waterfowl seeking breeding habitats are not attracted to the Kansas City area, despite the fact that the area is directly within the Mississippi Flyway. The Canada goose, common mallard, and blue-winged teal all prefer marshy to dry areas surrounded by relatively flat prairie vegetation (Johnsgard 1979). The Kansas City area is too hilly and forested to provide suitable nesting areas. In prehistory the floodplain and upland climax forests may have been considerably more extensive than what is observed today, thereby reducing the preferred habitat by an even greater amount.

Lack of fish and waterfowl as reliable resources to Kansas City Hopewell populations is directly reflected in the faunal remains. Fish bones comprise a very small percent (if any) of the assemblages from the Trowbridge (E. Johnson 1975), Young (Adair 1977), Yeo (O'Brien 1982) and Aker (Brown 1975b) sites. Likewise, migratory waterfowl are totally absent from these assemblages (except for one duck element identified from at the Aker site) with the Aves category being defined primarily by the remains of wild turkey (*Meleagris gallopavo*). The marginal exploitation of aquatic resources may also account for the low frequencies of bladelets at these sites (Reid 1976:72). Bladelets are tools whose morphological characteristics may be suited for skinning and butchering birds, scaling fish and working bird bone and shell. They represent one of the major tool types in Middle Woodland sites in the Illinois Valley where fish and bird remains are abundant (Parmalee et al. 1972).

Areas surrounding the Fisher-Gabert site in central Missouri are more attractive to migratory birds while the Missouri River at this location supports large populations of fish. These conditions are reflected in the recovered faunal remains from the Fisher-Gabert site. Included are elements of duck (*Anas platyrhynchos*) and Canada goose (*Grus canadensis*) along with a large number of fish bones. Identifiable remains include mainly catfish and gar (Brown 1975a).

Since the Kansas City Hopewell populations apparently did not rely on aquatic resources as major subsistence items, it is suggested that the sedentary lifestyle was maintained by a continual and steady exploitation of select plant and animal species; a strategy that may have required periodic adjustments in the settlement system so that this exploitation strategy could be retained. Analyses of floral and faunal remains from the Young site suggest that the preferred animal species, the white-tailed deer, was hunted according to a well defined seasonal procurement strategy. Not only did such a scheduling provide the inhabitants with prime meat, marrow, and hides, but it proved beneficial to the maintenance of the deer population as well (Adair 1977:68-79). In an analysis of 37 trash filled storage pits from the Trowbridge site, Bell (1976) demonstrated a pattern of economic stability in which deer were hunted and butchered according to a stable and consistent pattern for a period of about 300 years. The selection of upland climax forest products, especially hickory (*Carya spp.*) and hazel (*Corylus sp.*) provided, along with animal meat, the most concentrated sources of protein and calories. Remains of these plant foods are abundant at Kansas City Hopewell sites and are indicative of a successful collection strategy. The desire to retain such a complex pattern based on scheduling according to seasonal diversity may have been a prime factor in *both* the establishment of small ancillary habitations in close proximity to large permanent settlements and the incorporation of cultigens into the subsistence realm.

Johnson (1976, 1979) has suggested that small Kansas City Hopewell sites, such as Deister and Yeo, were established as either specialized procurement stations for resources necessary for the maintenance of larger villages or as daughter communities designed to compensate for overpopulation. While these hypotheses have not been empirically tested, both suggest that small camps were necessary establishments that alleviated pressure at the larger villages; a pressure implied to be caused by an imbalance between resources and human populations. For example, at Deister the artifact categories are indicative of fewer activities than observed at the Renner site (a large village perhaps affiliated with Deister) and are focused primarily on hunting, chipped stone tool manufacture, and food processing (Katz 1974). Additionally, most small camps are located upstream from larger villages (A. Johnson 1974), in closer proximity to upland climax forest products. The placement of these sites in such an ecological zone was probably related to the Kansas City Hopewell's heavy reliance on nut masts. Ford (1979:236) has suggested that the nut bearing trees utilized by Middle Woodland populations produced a more dependable crop in the upland zones than the same species growing in the floodplain forests adjacent to the large villages; therefore the climax forest products were reliable resources for the Kansas City Hopewell as long as mobility was possible. In a semipermanent or sedentary settlement pattern, mobility for resources is provided by smaller portions of the group traveling to the specific areas for the desired resources (Jochim 1981:149-155). Establishing smaller camps in the Kansas City region allowed for the continuation of the existing hunting and gathering strategy and helped maintain the economic stability noted from Trowbridge. It should be realized however that the location of settlements may have included factors other than those caused by a disequilibrium of food to people; factors of social and political organization may have also influenced the settlement pattern (Asch et al. 1979).

It has also been suggested that population expansion in the Kansas City region necessitated the use of cultigens as an added food supply (A. Johnson 1979), an alternative recognized as a possible strategy as resources and total population become strained (Ford 1979). It is argued however that these foods were not incorporated into the diet as a *change* in the economy, but as a selection on the existing subsistence base. Table 4.2 demonstrates the comparable nutritional qualities of the cultigens to other food remains present in Kansas City Hopewell sites. Oily seeds, which are high in calories, protein, and fat content, contain small percentages of water. These qualities, present in *Iva*, *Helianthus*, and *Cucurbita* seeds, are similar to those observed in the nut masts, particularly walnut and hickory. Starchy seeds, as their name implies, are high in carbohydrates and calories and relatively low in protein and fat content. Maize is very similar to the *Chenopodium* and *Polygonum* in this regard. Fruits are represented by grape, cherry, persimmon, blackberry or raspberry, and *Cucurbita* flesh. All of these foods contain high amounts of water which keeps the calories, protein, and fat at low levels. These foods are excellent for drying and storage for winter consumption. In fact, after the water content has been reduced by drying, the nutritional qualities of the fruit increases per weight.

Similarities can also be drawn between the cultigens and wild flora in terms of skills needed to process the food and the season of availability which dictated the collecting of the food. Marshelder and sunflower both contain seeds enclosed within a pericarp which is often removed before consumption. The technology needed to accomplish this task is extremely simple and can often be done without specialized prepared tools. Large numbers of these seeds could be processed using grinding slabs, boiling in water, or parching over a fire until the pericarp separates from the seed. These techniques were certainly not newly acquired skills; in fact, shelling a black walnut probably required greater strength and a special tool for prehistoric populations (Asch et al. 1972). Cucurbita seeds and flesh do not require any processing, although the seeds are far more tasty after they have been roasted. Corn may have been the only cultigen which required a new technique since removing the kernels from the cob is not quite the same as extracting nut meat from the shell. However, it is difficult to imagine this being a complicated task since mature corn shells quite easily with a thin blunt tool. Ethnographic accounts describe various ways in which corn was processed. Wilson (1977;40-41,48) describes corn shelling among the Hidasta, a process that included the use of small pointed sticks, mussel shells, or just the thumb. The technique was determined by the purpose of the corn; seed corn was usually shelled with the thumb while dried corn required a tool. Occasionally corn did not require any shelling since it was eaten directly off the cob after it had been parched or boiled. Will and Hyde (1964:147-164) discuss ways in which corn was prepared by various tribes in the Northern and Central Plains; included are boiling, parching, drying, and grinding into flour. All of these methods are considered as likely skills for any group who included plants, cultivated or wild, as a major contributor to their diet.

Most seeds are mature and available for harvest in the summer and fall months, the exact time being dependent upon the date of the last frost in the spring and the length of time required for the particular seed to ripen. Berries are some of the earliest plants to produce mature fruit and are generally available in mid-summer. Present day varieties of squash and gourd require 50 to 120 days to mature and the early varieties, the summer squashes, are also ready for harvest in mid-summer. The winter squashes, pumpkins, and gourds are mature in early fall. Small seeded annuals such as *Chenopodium*, *Helianthus*, *Polygonum*, and *Iva* ripen in the late summer (Steyermark 1963), although the sunflower can be harvested as late as after the first frost. Nuts are generally fall crops (September and October) although the hazel matures earlier, beginning in August and continuing through September (Asch et al. 1972). Corn, if planted in late spring, is mature in late summer (August to September).

These similarities suggest that certain cultigens were added because they fit into pre-existing categories of food and supplied diversity to the diet without changing nutritional demands or techno-

Table 4.2. Proximate Composition of Vegetal Foods Utilized by Kansas City Hopewell Populations.

Plant	*Food Energy Cal/100 g*	*% Water*	*% Protein*	*% Fat*	*% Total Carbohydrate*	*Calcium*	*Phorphorus*	*Iron*
Oily Seeds								
marshelder kernels *Iva annua*[a]	535	5.06	32.25	44.47	10.96			
sunflower kernels *Helianthus annuus*[b]	560	4.80	24.00	47.30	19.90	120	837	7.1
pumpkin and squash seeds *Cucurbita*[b]	553	4.40	29.00	46.70	15.00	51	1144	11.2
Starchy Seeds								
goosefoot *Chenopodium spp.*[b]	320	13.40	13.30	5.60	45.90			
field corn kernels *Zea mays*[b]	348	13.80	8.90	3.90	70.20	22	268	2.1
knotweed								
Polygonum pensylvanicum[d]			9.90					
P. convolvulus[d]			14.40					
Fruits								
blackberry *Rubus ostryifolius*[b]	58	84.50	1.20	.90	12.90	32	19	.9
raspberry *Rubus occidentalis*[b]	73	80.80	1.50	1.40	15.70	30	22	.9
cherry *Prunus americana*[b]	58	83.70	1.20	.30	14.30	22	19	.4
grape *Vitis sp.*[b]	70	81.90	1.40	1.40	14.90	17	21	.6
persimmon *Diospyros virginiana*[b]	78	78.20	.80	.40	20.00	6	26	.3
pumpkin flesh *Cucurbita pepo*[b]	26	91.60	1.00	.10	6.50	21	44	.8
squash flesh *Cucurbita sp.*[b]	19	94.00	1.10	.10	4.20	28	29	.4
Nuts								
black walnut *Juglans nigra*[b]	628	3.10	20.50	59.30	14.80	T	570	6.0
pecan *Carya illinoensis*[b]	687	3.40	9.50	73.70	12.70			
hickory *Carya ovata*[a]	696	2.20	11.00	72.70	10.60			
hazel *Corylus americana*[c]	630	2.60	22.50	61.40	11.20			
white oak *Quercus alba*[c]	221	47.30	2.80	3.30	43.90			
red oak *Quercus rubra*[c]	229	38.20	5.60	20.80	68.10			

[a] Asch and Asch 1978; [b] Watt and Merrill 1963; [c] Wainio and Forbes 1941; [d] Earle and Jones 1962

logical skills in collecting, extracting, and processing. These plants, however, were not staples. Other than the fact that their remains are simply not abundant in Middle Woodland contexts, from either Kansas City, Illinois Valley, or Ohio, three other lines of evidence may be supportive of this suggestion.

First, since corn becomes a dominant staple in later periods, it is viewed as the most likely of all Kansas City Hopewell cultigens to increase in importance during this time. The corn from the Trowbridge site however, as well as samples from sites to the east, show a wide range of genetic expression (Ford 1979:237). This variation tends to lessen predictability about yields and inhibits the crop from becoming a dependable resource. Uniformity in phenotypic and genotypic expression is associated with well-tended and carefully maintained maize varieties and indicates a product grown for high productivity rather than for seasonal security (Ford 1979).

Second, there was no major competition from other cultural groups for the catchment areas or resource zones utilized by the Kansas City Hopewell. If, in fact, smaller sites were fissioned off daughter communities, their location was not dictated by boundaries imposed by other cultures. Although most of the expansion occurred in the Kansas City region, actually creating a "packing situation" in the valleys, the presence of Kansas City Hopewell sites as far west as Manhattan, Kansas indicates the population had ample room to expand whenever population growth threatened to shrink the sizes of the individual site's resource zones. Additionally, initial data from the Elliott site (14GE312) in east-central Kansas, suggest that long range moves did not result in major changes in the subsistence strategy (see Chapter 3).

Third, trace element analyses of human bone from Illinois Valley burials indicate that animals were the primary sources of protein for these Middle Woodland populations (Buikstra 1979; Ford 1979). If such results can be applied to the Kansas City Hopewell, it is suggested that this implied preference for hunting was intensified following the introduction or increased efficiency of the bow and arrow during the latter part of the Kansas City Hopewell sequence. A general decrease in size of projectile points, particularly thickness, is noted during the entire Middle Woodland period (Montet-White 1968), although the final reduction in size from dart to arrow takes place rapidly from about A.D. 700-800. This new technology has been suggested as one of several factors connected with the habitation and corresponding resource exploitation of the upland communities during the late Late Woodland in Illinois (Asch et al. 1979; Styles 1981).

To summarize the situation during the Middle Woodland period, the following points are involved. First, domesticated plants, including corn, squash, and *Iva* were grown by the prehistoric populations. They were not staple resources however, but combined with others to provide seasonal security. This strategy can be extremely beneficial to a group whose resource base includes nut masts. Since nuts vary greatly in their yield and infrequently have two successive masts of measurable quantity, they are not a dependable food for resource security. Second, there was an apparent emphasis on hunting as a means of acquiring necessary energy and protein levels, although gathering of plant foods was also crucial to the subsistence strategy. This emphasis may have been more pronounced in the Kansas City region since the area does not attract a reliable quantity of fish or migratory birds. Third, as with any cultural group practicing a semisedentary existence within a defined area, population density may have caused an imbalance between available resources and group size. This problem may have been temporarily and periodically alleviated by either adjustments in group size and/or settlement location. An increase in the availability of domesticated plants was not a high ranking alternative since ecological, demographic, and technological conditions, and probably social constraints as well (Ford 1979), favored the continuation of the existing system.

Late Woodland—Kansas City Locality

As discussed in Chapter 2, only one single component Late Woodland site in the Kansas City area has been extensively investigated (K. Brown 1981). Recovered floral remains at Sperry (23JA85) were not abundant and did not include cultigens. This record, in conjunction with the presence of small, triangular notched arrow points, the lack of house remains, and the small size of the site, indicates a settlement-subsistence pattern that focused on hunting; a practice that involved smaller and perhaps more mobile groups of people than known for the preceding Middle Woodland period. If the Kansas City Hopewell did in fact emphasize hunting over gardening, then the addition of the bow and arrow to the technological makeup made this emphasis all the more attractive for the Late Woodland populations and may have led to a settlement-subsistence pattern witnessed at the Sperry site. To date, no large sedentary Late Woodland villages have been located to suggest that small sites such as Sperry were ancillary hunting stations designed to maintain the permanent settlements. While such a description differs from Late Woodland sites in Illinois, caution should be taken in assuming this one site is representative of the Late Woodland cultural period. Johnson (personal communication) calls attention to the presence of Late Woodland house sites along tributary streams of the Missouri River in Platte and Clay counties, while noting the lack of professional surveying and testing of these sites in the area. Martin (1976) also discusses the presence of small Late Woodland habitation sites in the Fishing River drainage just to the east of the study region. Most recently Schmits (1986) has identified two late Woodland phases in the Little Blue Valley and suggests a marked break in the cultural record between these phases and the earlier Middle Woodland occupations. Additionally, the two multicomponent Late Woodland-Steed Kisker sites in the Smithville Lake area (McHugh 1982) contained cultigens and one demonstrated remains of a structure. McHugh (1982: 174-177) argues however that this structure is associated with the Steed-Kisker occupation. Until a considerable amount of surveying and excavation has been conducted, the extent and nature of the prehistoric population in the Kansas City area between A.D. 500-1000 will remain unclear, as will the importance of domesticated plants. However, if the Late Woodland period in the Kansas City region shows any similarity to what will be presented for Woodland cultures to the west, we should expect this cultural period to represent semi-sedentary populations who maintained and dispersed domesticated crops and who were slowly evolving towards a dependence on these foods recognized as an agricultural system.

Middle Ceramic Agricultural Systems

The botanical data throughout the study area demonstrate the presence of agricultural systems by about A.D.

1000. Such systems are initially recognized in northcentral Kansas (Solomon River phase), southcentral Kansas (Bemis Creek phase), and the Kansas City locality (Steed-Kisker phase). According to the proposed model, these Middle Ceramic period systems developed out of domesticatory systems maintained by previous populations. These preceding groups may be conceived of as either cultures already living in the Central Plains or as populations migrating in from outlying areas. This latter suggestion has been presented as a possible origin for the Central Plains tradition (Wedel 1959, 1961) and deserves some consideration here.

Ceramic designs, point styles, and permanent earthlodge structures of the Central Plains tradition exhibit some similarities to cultures located south and east of the study area. These similarities have prompted several studies, including a trend surface analysis of 112 radiocarbon dates from 58 sites throughout the Central Plains (Roper 1976). Although Roper (1976) was not arguing an origin for the Central Plains tradition, results of this analysis indicate a general south to north trend in dates. Such a trend may be reflective of an expansion route of populations who may have been agriculturalists. This might explain the sudden appearance of cultivated plants after an absence of several hundred years (See Fig. 3.9).

While population movement throughout the Plains and surrounding areas may have significantly influenced the demographic patterns of the study area, several reasons suggest that it does not completely account for the origins of agricultural systems. First, the trend surface analysis explains only 21 percent of the variance in the radiocarbon dates when both latitude and longitude were used for prediction. Second, the analysis only accounts for existing Central Plains sites with radiocarbon dates and does not include any late Plains Woodland sites that are relatively dated from A.D. 700-1000. Third, dates from two cultural complexes, Solomon River and Steed-Kisker, do not fit the chronological evidence of a south to north route; these complexes are two of the three that provide botanical evidence of an initial reliance on domesticated plants. Fourth, while some defined Middle Ceramic complexes apparently exhibit a south to north movement, other studies indicate their development was due to a local transition from earlier indigenous Woodland complexes (Adair and Brown 1981; Benn 1981; Jantz et al. 1978; Steinacher 1976). Fifth, although the data are not as definite as one would like, cultigens have been recovered from some late Plains Woodland complexes, at least one of which spans the transition from Woodland to Village periods. Clearly, the issue needs further attention, which is beyond the scope of this study, although present data indicate that the development of agricultural systems cannot be explained directly by a migration of people for the entire Central Plains.

Migration/Diffusion

Although the status of domesticates in the Late Woodland phases in the Kansas City region is unclear, the Steed-Kisker phase provides some of the earliest evidence of agricultural systems in the same locality. Sufficient data are available to suggest that this particular phase may have been partially the result of Middle Mississippian populations who moved into the area from an eastern origin. It is suggested here that this phase and the corresponding agricultural system it maintained provides the only evidence that may be in part attributed to migration or diffusion. Exactly how much of the subsistence strategy was due to such a move is not known since Late Woodland populations in the Kansas City area may have possessed domesticated plants. Three lines of evidence however, suggest that a movement of peoples or information was at least partially responsible for the recognition of the Steed-Kisker complex.

First, O'Brien (1974, 1976, 1978a) has repeatedly demonstrated the similarities in ceramic designs between Steed-Kisker and Middle Mississippian sites to the east. Additionally, these designs (plus shell temper and vessel form) are only minimally represented in other Middle Ceramic complexes in the Central Plains. Plus, as presented in Chapter 2, the Steed-Kisker ceramics show no relationship with the previous local Late Woodland variety.

Second, analyses of skeletal remains (primarily cranial measurements) from the Calovich mound (Barnes 1977) and the Steed-Kisker cemetery (Stewart 1943) indicate a population biologically distinct from other populations in the Central Plains. Some biological ties are noted however between the Calovich and Sugar Creek (Nebraska variant) populations, suggestive of some affinity. Interestingly, Barnes (1977) notes similarities in mortality curves, mortuary offerings, and burial treatment of the Calovich materials to Dickson Mound (Middle Mississippian) in Illinois but demonstrates that they differ biologically. This conclusion may suggest that the Steed-Kisker phase was not entirely due to a direct migration between the Mississippi River Valley and the Kansas City locality, but a gradual movement that biologically involved the incorporation of local peoples.

The third line of evidence is provided by the botanical record. As previously discussed, Northern Flint is a distinctive but variable and rather late type of corn and is most common in prehistoric times in New England, Upper Missouri, and Middle Mississippi area sites. The only recognition of Northern Flint maize in the study area is from several Steed-Kisker sites (Cutler and Blake 1973b). This variety was also identified from Cahokia although it was not dominant (Cutler and Blake 1973a). The presence of Northern Flint in only Steed-Kisker sites suggests that this cultural complex maintained some relationship with populations to the East where the variety may have actually developed (Yarnell 1964). An early historic context for the Northern Flint variety in the study region comes from the Kings Hill site (23BN1), a Kansas occupation in northwest Missouri (Blake and Cutler 1982).

Internal Developments

If agricultural systems in areas other than the Kansas City region cannot be accounted for by an in-migration or "diffusion", then they must logically be attributed to developments out of indigenous domesticatory systems of the Plains Woodland period. In addition to the botanical data already presented, two other lines of evidence support the likelihood of such a system in the study area prior to A.D. 1000.

The influence of the Kansas City Hopewell on the development of Plains Woodland complexes is not clearly known although the latter phases represent distinctive changes in ceramics, settlement sizes, and burial practices from the former. As far as domesticated plants are concerned, it is hard to imagine squash, corn, and *Iva* cultivation being re-introduced into the Central Plains after the decline of the Hopewell. Rather, it is easier to conceptualize, particularly for the Kansas City area, a situation where these domesticates were retained by later populations and gradually increased in importance. The dispersal of Kansas City Hopewell hunting parties or complexes such as Schultz and Cuesta may have been critical in the spread of cultigens throughout a good portion of the study area and thus may explain the availability of these foods to Plains Woodland populations. Sunflower cakes have been described by Wilson (1977) as excellent sources of protein and energy used by traveling groups or hunting parties. Given the nutritional similarities between sunflower and marshelder, it is possible that the latter plant served a similar purpose. The single *Iva* achene from the Elliott site and the two maize kernels from the Cuesta phase may indicate that the Kansas City Hopewell did in fact disperse cultigens and/or horticulture as they moved throughout the area or as they influenced existing populations. How extensive this dispersal was, both in terms of the number of cultigens and in the spatial distribution of these foods cannot be determined with so little data.

Second, emerging data on sites temporally placed in the latter part of the Plains Woodland period (A.D. 700-1000) indicate that populations were organized into small semi-sedentary village communities along waterways throughout the study area (Brogan 1982; Grange 1980; A. Johnson 1968; Reynolds 1979; Leaf, personal communication). Excavations and analyses consistently reveal one or several small (5-7 meters in diameter) structures that may represent a single or several nuclear families living and traveling together and perhaps continually re-occupying the same locations. The degree of sedentism is unknown although judging from the size of the structural posts and often the lack of an interior hearth, year round occupation seems unwarranted. If the number of houses, habitation sites, and the amount of cultural debris are any indication of population size, a relatively extensive population is indicated. Tools recovered from these sites include grinding slabs and manos, indicative of the use and processing of vegetal foods, but as yet, no agricultural implements are associated with this time period. The picture that emerges is similar in many respects to the Kansas City Hopewell, suggesting that cultigens grown for varietal or security reasons would have been quite consistent with the lifestyle of the Plains Woodland people.

While this topic is the focus of the following chapter, it is important to stress here that the question of plant utilization during the Plains Woodland period is a problem not restricted to just cultigens but to floral remains in general (See Table 2.5). Considering our poor understanding of the species of plants represented in Woodland deposits, it is not surprising that the relative importance of cultigens is unknown. This problem may be attributed to two primary causes. First, Plains Woodland sites are often located partially within the plow zone and are subject to various disturbances, including displacement of artifacts due to farming, soil erosion, and different bioturbations. These disturbances often leave the prehistoric occupations "fuzzy"; house structures are not clearly defined, features can be extremely difficult to recognize, and intact middens are sometimes tenuous at best. Second, excavation strategies designed for the recovery of floral remains focus primarily on the flotation of feature matrices. In situations where features are poorly defined or truncated due to farming practices, recovered plant remains are usually modern contaminants. These problems clearly need attention if the subsistence strategies of Plains Woodland populations are ever to be properly understood.

At this point, an inference is made from the combined evidence that cultigens were present and grown by Plains Woodland peoples, but their actual importance awaits future research. The issue now is to explain why the availability of domesticated plants was increased by prehistoric populations, or in essence, to explain why agricultural systems developed. The discussion that follows can be divided into two sections, one which focuses on general trends evident in the archaeological record throughout the study area, and a second which focuses on specific areas or localities where botanical data are sufficient to demonstrate well developed agricultural systems.

General Trends

Data accumulated throughout the study area, (although research has concentrated more on the eastern half), suggest that by the latter part of the Woodland period (A.D. 700-1000), semi-sedentary populations were economically maintained by a significant reliance on river valley resources. Upland hunting or gathering is not denied in this scenario, but the valleys provided the reliable and dependable resources needed to sustain populations who had gradually become more sedentary. Included under the distinction as reliable and dependable resources are fish, migratory birds, and plants (Flannery 1968; Jochim 1981: 138). Other resources, specifically large mammals, are at times less reliable since their adaptability allows them to occupy environmental niches other than just river valleys. Subsequently, they must be tracked, sought after, and hunted, while the alluvial valleys maintain high biomass levels due to the more restricted adaptability of the resources (Smith 1978).

Through time, and most evident with a sedentary lifestyle, populations in the river valleys most likely increased. The absolute magnitude of this population increase cannot be determined without sufficient demographic parameters, although the direction of change was consistent in that populations were apparently increasing. This trend posed no threat to the economic stability of prehistoric groups as long as sufficient room was available for dispersal or fission or as long as cultural practices (abortion, infanticide, etc.) maintained the group to a size compatible with the available resources. In the Central Plains, the data indicate that populations gradually expanded to the point that nearly all major river valleys were occupied. Increasing population densities and sedentism probably occurred concurrently and may have also included an increased use of cultigens. Survey records from areas in eastern Kansas (Logan 1981, 1983) suggest that Woodland sites along major rivers have been destroyed or deeply buried by frequent and severe floods and changing channel courses so that most Woodland sites found today are along tributary streams. This may indicate a far greater population level than the density of known sites suggests.

An extensive occupation of the valley areas created a "population packing" situation (Bronson 1977), or what might be referred to as pseudo-population increase. Ample room was available in the uplands for occupation, but the economic demands of the population could not be satisfied in these zones. When expansion was no longer a viable alternative, that is, when expansion and economic stability were no longer compatible, a pressure situation was created. The actual number of people involved was not critical to the situation; there were simply too many people of the same cultural adaptation wanting the same resources from the same micro-environments.

Throughout the latter part of the

Woodland period, subtle pressure situations were created but were never catastrophic occurrences. An increase in the availability of domesticated plants may have been a high ranking alternative for prehistoric populations whenever pressure situations were recognized. Again, in a suggestion similar to that presented for the Kansas City Hopewell, an increase in the consumption of agricultural foods was not an attempt to *change* the economic pattern but rather an attempt to *retain* it. Four lines of evidence are presented as a support to this suggestion and also provide evidence to argue against the development of agricultural systems as being predominantly a direct result of population increase.

First, the botanical data clearly demonstrate that domesticated plants were grown, consumed, and maintained by populations who did not depend on them for survival prior to the emergence of agricultural systems. While the data are not as abundant for some time periods in the Central Plains, similar trends have been identified from other areas throughout the United States (Asch and Asch 1982; Ford 1981). The presence of cultigens in a domesticatory system is attributed to their compatibility in nutritional qualities, processing techniques, and gathering periods with a variety of native flora utilized by the prehistoric inhabitants. By A.D. 1000, these foods were well known and well accepted by populations throughout the Central Plains; they did not suddenly appear when people needed an alternative to increase resources.

Second, significant genetic changes had occurred, especially in maize, that rendered the plant a more environmentally reliable and productive resource. Included was an increase in phenotypic and genotypic homogeneity of the major variety, although different varieties continued to be grown until historic times.

Third, the plants' productive qualities and the longevity of their acceptance as general food items were important cognitive factors that guided the prehistoric populations in their activities to increase the availability of the cultigens. This activity was not simply a response to environmental constraints, for the decision of what to eat, how to prepare it, and with whom to share it are all part of a cultural code (Wetterstrom 1978), which is something not likely to change overnight. Human populations were directly responsible for increasing the ecological niche of cultigens and they did so at the expense of the habitats of the natural resources. In most areas, the floodplain forest habitats which were altered were the same ones that had provided the prehistoric inhabitants with the annually renewable resources that once constituted significant portions of the food base. This action was of greater consequence in the western portion of the study area where gallery forests are narrow and very dependent upon the moisture from the river for survival. Such a measure would not have been taken if prehistoric populations had not been aware of the reliability and dependability of cultivated crops. The decision to increase the availability of cultigens was primarily cultural and was related to existing parameters of social organization, division of labor, and technological skills. Changes in these institutions are relatively gradual processes in their own right and if related to subsistence strategies, they developed along with agricultural systems. If the botanical data base is strong and sound enough to indicate that agricultural systems actually developed in the Central Plains from the Woodland subsistence strategy, then changes in related institutions should also be preserved in the archaeological record. Such changes have been methodologically argued from surrounding areas (Braun 1977, 1981; Braun and Plog 1982; Plog 1974). In the Central Plains, agricultural systems are accompanied by larger, perhaps extended family, structures; organized village communities; and regional conformities in ceramic styles.

The reliability of domesticated plant resources may have been importantly influenced by climatic change. Geomorphic formations and soil horizons indicate a period of stability or equilibrium had been attained by approximately 2000 B.P. in northcentral Kansas and by 1000 B.P. in southcentral Kansas. Artz (1983) has argued that such conditions were the effects of changing climatic patterns, specifically the duration, intensity, and frequency of thunderstorms that resulted in periods of valley degradation and aggregation. Data from the lower Walnut Valley in southcentral Kansas demonstrate that from approximately 4500-2000 B.P. the area experienced a greater frequency of moderate magnitude floods, perhaps triggered by intense spring or fall frontal storms. During the following period, 2000-1850 B.P. the frequency of large floods decreased, stream channels widened and deepened, upland erosion was evident, and short-term severe droughts may have occurred. The following episode, 1850 B.P.-present marked the "development and maintenance of a geomorphic adjustment to the present hydrologic regime" (Artz 1983:254). In terms of prehistoric groups inhabitating the valley areas, greater reliability and predictability of many climate-dependent resources could have induced a more intense occupation and perhaps the adoption of agriculture during the last episode. Artz's study has significant implications for most regions throughout the Central Plains where prehistoric cultivation was possible only in the river valleys since the average amount of rainfall is inadequate for crops grown in the uplands without some form of irrigation.

Specific Localities

In parts of northcentral Kansas occupied by the Solomon River phase populations, agriculture was possible only in the valleys where the water supply was adequate and constant, where the alluvial soils were deep and well developed, and where high bluffs or buttes protected the valleys from strong prairie winds. Wedel (1941:4) describes the area as:

". . . the eastern front of the High Plains (which) has been dissected by stream erosion to produce the so-called Plains Border. This is a belt of high plains and prominent east-facing escarpments cut through by the deep, broad valleys of the Smoky Hill, Saline, Solomon and their tributaries. Buttes and ridges are a conspicuous feature of the landscape. The fertile river valleys are fringed with forests of oak, elm, ash, walnut, cottonwood, and other hardwoods, and bluestem and bunch grasses formerly clothed the upland areas."

Semi-sedentary populations recognized as the Keith variant may have increased the availability of cultivated plants given the demographic, botanical, cognitive, and environmental factors which favored such a subsistence strategy. Archaeologically, this change is reflected in the recognition of the Solomon River phase. Since the geomorphic discontinuities, observed in the soils and pollen columns of Deer and Wolf Creeks, tributaries of the Solomon River, indicate an earlier period of alluvial stability in this area than recognized in the Walnut River drainage (Artz 1983; W. Johnson 1981), northcentral Kansas may have been better suited for intense river valley occupation and cultivation during the mid to latter part of the Woodland period. Radiocarbon dates and botanical remains support this suggestion and may explain why the Solomon River phase does not fit Roper's (1976) trend surface analysis. Extensive exploitation of gallery forest zones has been noted for both the Keith variant (Grange 1980) and the Solomon River phase (Falk 1969b). Interestingly, Falk (1969b:51) has suggested that the area could not have sustained such an exploitation on an annual basis for an extended period of time if groups were permanent. Given the suggested population packing, an extensive use of native valley resources would have certainly caused periodic pressure situations in prehistory.

Some caution in regards to radiocarbon dates should be expressed here. While they do appear to support an earlier development of agricultural systems than do the dates from the Bemis Creek phase in southcentral Kansas, this suggestion may be unwarranted. The temporal placements for some Solomon River sites are based on a single radiocarbon date from one house. A much tighter temporal control can be gained by statistically averaging several dates from a single house (Hotopp 1978). The date of A.D. 1000 given for the Two Deer site is an average of six dates from one structure. The range of time represented by these dates is A.D. 860-1060 (without the sigmas), which is very similar to the Solomon River range of A.D. 830-1190.

Conditions were apparently favorable for intense cultivation in the river valleys of western Kansas and Nebraska during the early Middle Ceramic period. Upper Republican villages are found on almost every major stream and tributary in this area (Wood 1969b:103), representing the only known situation in which non-irrigated corn agriculture was pushed well beyond its present day western limit (Wedel 1961). Although many of the rivers and streams in these western areas are fed by natural springs, the relative uniformity in the intensity and frequency of thunderstorms, measured in geomorphic stability (Johnson 1981), may have been a factor in providing stable areas for alluvial field cultivation.

The agricultural system at the Two Deer site in the Walnut River drainage may also be explained by the culmination of demographic, botanical, cognitive, and environmental factors. Forthcoming studies (Gary Leaf, personal communication) will demonstrate the sedentary valley oriented characteristics of the preceding Woodland Butler phase, while Artz (1983) has focused primarily on this area to provide empirical evidence of climate induced geomorphic conditions. Additionally, archaeological investigations have indicated strong similarities among the Woodland sites excavated, suggestive of a single cultural phase. Radiocarbon dates from these sites agree with the onset of stability in the valleys at approximately A.D. 700-750. Although inconclusive in itself, the single maize kernel from 14BU57 suggests the presence of cultigens in a hunting and gathering strategy before the recognition of agricultural systems.

Steed-Kisker agriculture has already been presented as possibly being influenced by migration into an area occupied by semi-sedentary peoples. The reliance on domesticates, especially maize, during the Steed-Kisker phase was probably greater than the floral remains indicate. Analyses of the pathologies on human skeletal remains from the Calovich burial mound (14WY7) provide evidence suggestive of a pronounced consumption of maize by Steed-Kisker populations. Barnes (1977) reports the presence of porotic and orbital hyperostosis in 26 percent of the population. This condition refers to a number of bony lesions found mainly on the superior border of the orbits or on the external surface of the cranial vault. Some feel it is one of the best signs of nutritional imbalance in the archaeological record. The condition is caused by severe anemia and may be associated with chronic dietary iron deficiency (El-Najjar and Robertson 1976) or hookworm (Angel 1975) in the New World. Several studies (Lallo et al. 1977; El-Najjar and Robertson 1976) indicate that the frequency of porotic hyperostosis is significantly higher in agricultural communities than in hunting-gathering groups, particularly those dependent on maize agriculture. In addition to being low in iron (see Table 4.2), maize contains phytate which binds iron and reduces the amount absorbed by the biological system. Well known alkali cooking techniques (the addition of lime to cooking water or fires), as described by Wing and Brown (1979) and Katz et al. (1974), free otherwise unavailable iron and essential amino acids in corn. The use of such techniques in prehistory is unknown in the study area, although Braun (1981) has suggested that during the Late Woodland period in the Illinois Valley, ceramic forms and construction favored the long boiling of starchy foods. While excessive boiling in water destroys most vitamins and minerals, the phytate content can be reduced during processing, thus allowing for the increased absorption of iron (Bender 1978; Wing and Brown 1979).

Tooth decay (i.e. caries) is closely associated with sugar content and consistency of food, occurring at high frequencies in high carbohydrate diets (Cassidy 1980), such as maize. Caries were common in both deciduous and permanent dentition of individuals from Sugar Creek ossuary (Nickels 1971:43-45), Calovich mound (Barnes 1977:149-152), and Steed-Kisker cemetery (Stewart 1943:254). They range from pin-point to large carious lesions and are observed on all age groups, but most pronounced in individuals over 40 years.

Factors involved in the development of agricultural systems therefore varied between the three areas where such a system is first recognized. A major difference between the Walnut Valley and the two other areas focuses on developments during the Middle Ceramic period. In northcentral Kansas the Upper Republican variant is represented by sedentary agricultural villages until A.D. 1250-1300, while the Nebraska variant of northeast Kansas/northwest Missouri continues until A.D. 1400. Occupation in the Walnut Valley declines significantly after the Bemis Creek phase with very few sites providing diagnostic or radiocarbon evidence of Village period residence. Tests at a single site of this period, 14BU71, recovered the only evidence of domestication; one sunflower achene. The relative abandonment of the Walnut Valley after A.D. 1100 may have been related to social or political factors rather than to economics. The intensity of village sites with agricultural systems appears to be located to the west and south at this time; specifically along the Bluff Creek drainage.

To summarize this chapter, prehistoric cultivation in the Central Plains subarea is identified by two different systems which occurred at two separate periods separated by a short temporal and/or spatial hiatus. The first, which may be classified as a domesticatory system, was associated with the Kansas City Hopewell. The second, recognized as an agricultural system is initially observed in three localities in the study area: northcentral Kansas (Solomon River phase), southcentral Kansas (Bemis Creek phase), and the Kansas

City area (Steed-Kisker phase). Botanical, ceramic, and skeletal data suggest the agricultural system in the Kansas City area may have been partially due to a migration of people or ideas into the region. The observed subsistence systems in the other two areas resulted from an indigenous development out of preceding Woodland subsistence systems. Factors involved in this development may have included general population increase and packing in valley areas; gradual trends toward sedentism; complements in nutritional qualities, seasonal availability, and collecting and processing techniques between the cultigens and several heavily utilized native resources; positive genetic changes in maize; and a cognitive awareness that certain plants could provide a dependable and necessary food source if their availability could be increased. Environmental stability may have supported the shift towards increasing use of cultigens, but was not a single limiting factor in the gradual process towards an agricultural system.

CHAPTER 5

The Spatial Distribution of Cultigens: A Case Study from South-Central Kansas

This study has focused primarily on the appearance of domesticated plants as a basis from which to discuss changing subsistence strategies in the Central Plains. It was argued in the preceding chapter that Plains Woodland populations who inhabited the Central Plains between A.D. 500-1000, contributed to the development of Middle Ceramic agricultural systems by growing, maintaining, and dispersing cultigens. At present, however, the botanical data from Plains Woodland sites provide a weak support for this argument. In this chapter, the distribution of identified cultigens from within a house structure at the Two Deer site (14BU55) is presented as a case study which may suggest a major reason for the lack of identified cultigens in late Plains Woodland sites throughout the study area.

Two Deer is a multi-component occupation located along the west bank of Bemis Creek, a tributary of the Walnut River in Butler County, Kansas. Surface diagnostics indicate Archaic, Woodland, and Village period occupations, although distributions of these artifacts suggest spatially separated localities (Fig. 5.1). Initial test excavations in 1975 (Fulmer 1977) and extensive salvage excavations in 1978 - 1979 (Adair and Brown 1981; Adair 1982) focused on approximately 212 m^2 of a single occupation manifested in the remains of several structures and dated to A.D. 1000 ± 25. The material culture remains from these excavations were used to formulate the Bemis Creek phase (Adair 1983), a transitional Plains Woodland/Plains Village complex in which major developments such as a sedentary lifestyle and an agricultural system are recognized.

Chipped stone tools include side-, corner-, and un-notched projectile points; small, thin oval bifaces; beveled knives; large celts and choppers; various edge utilized scrapers; hammerstones; perforators; notches; cores and lithic manufacturing debris; and numerous retouched and utilized flakes. Worked stone implements include sandstone abraders, nutting stones, and large grinding slabs or metates. The ceramics are quite abundant and distinctive; dominant characteristics being burned bone tempering; cordmarked exterior surfaces; straight rims with slight shoulders; and long bodies with a conoidal base. Variations from this description include sherd and clay temper; smooth, floated exterior surfaces; slightly flaring rims with more prominent shoulders; short bodies with an overall globular shape; and incised line decorations on the neck portion of the vessel. Most artifacts were confined within a 12-15 cm level and are reflective of a single occupation. Therefore, arbitrary 10 cm levels were combined for subsequent analyses. These studies demonstrated the presence of predominantly Plains Woodland artifacts with the introduction of lithic and ceramic styles more commonly associated with the Plains Village Tradition.

Two structures were recognized during excavations at Two Deer, although most attention was directed toward House 1. This structure was identified primarily by the concentration of burned daub since exterior post-mold stains were not visible in the field. Its estimated size is approximately 10 m along the

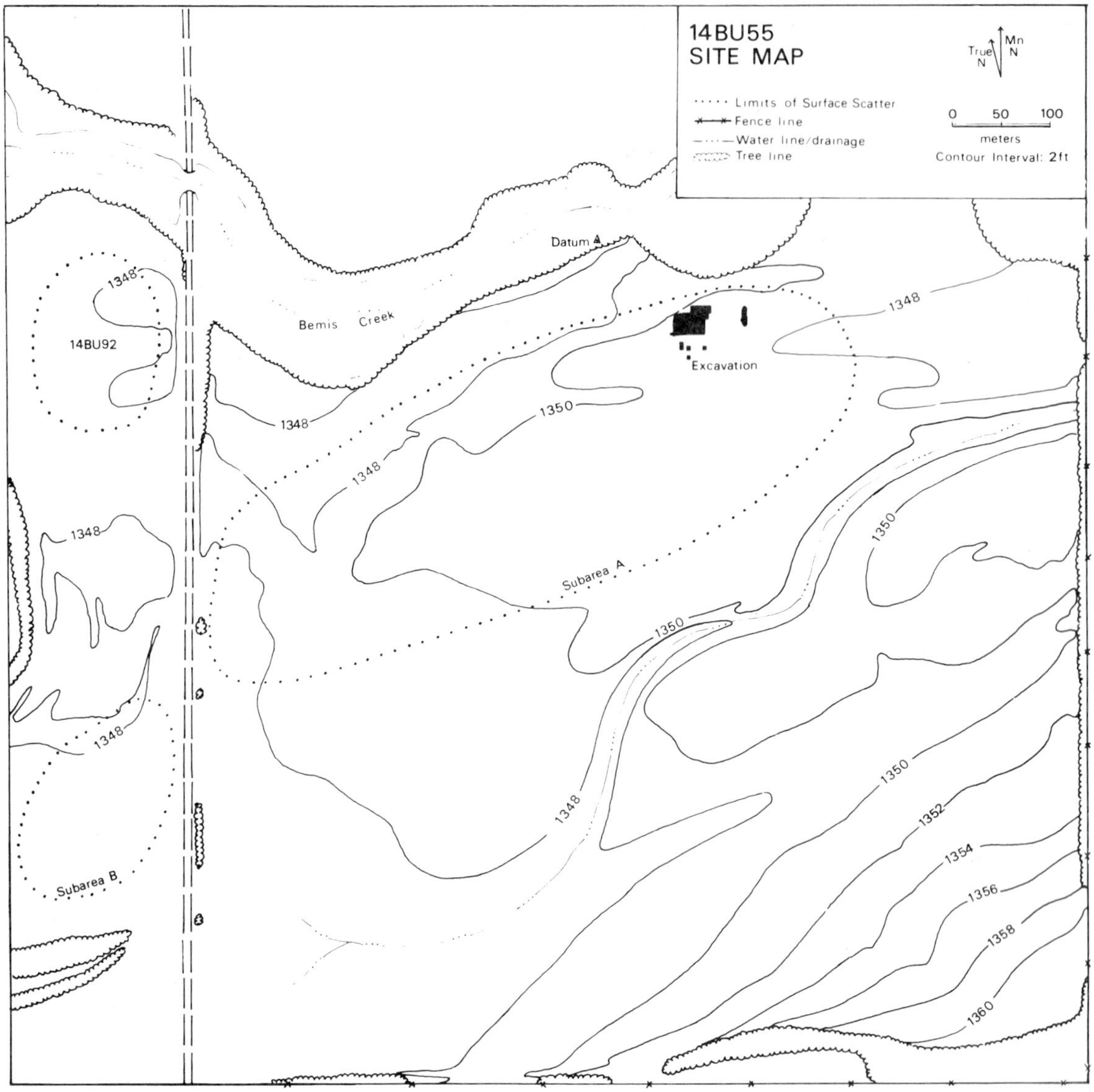

Fig. 5.1 Surface distribution of cultural materials and location of excavation, Two Deer site.

major axis (Fig. 5.2), although an exact outline of the shape and size cannot be determined. Postmold stains outline a portion of a second structure to the northeast of House 1. This second house has been largely destroyed by erosion, farming, and its location on the edge of the existing terrace. In fact, labeling the post stains as a house may actually be unwarranted. Over 90% of all recovered artifacts were located in the interior of House 1, which appears to have been abandoned after it burned and collapsed. The artifacts were dispersed throughout the house, although definite concentrations of ceramics and chipping debris were noted (Adair 1982). These concentrations were used to suggest that tool manufacture and food preparation activities were repeatedly carried on inside the structure around the centrally located hearth and metates (Fig. 5.2). The placement of activities around these features may have been related more to their central position as a focal point for the group or family (cf Yellen 1977) than to their function in the particular activity.

Identified faunal remains were limited but suggested an exploitation of adjacent floodplain forest and upland prairie environments for the utilization of deer, bison, raccoon, and beaver. Flotation of soil matrix was extensively used during excavations at Two Deer. Control samples of one-half bushel were taken from each excavation unit (lm x lm x 10cm) and feature fills were entirely floated. In some units, high charcoal concentrations prompted additional sampling. These strategies were successful in recovering almost 7000 charred seeds and 107 grams of nutshell fragments (Tables 5.1 and 5.2). While the presence of some of the carbonized flora may be due to accidental charring and fortuitous placement within the habitation area, the identified taxa indicate a heavy reliance on wild and cultivated plants by the prehistoric occupants.

The identified plants which most likely represent food resources are listed in Table 2.7 of this study (page 70). Included are ten wild annual plants or herbs, three nut bearing species, and four cultivated plants. Other identified

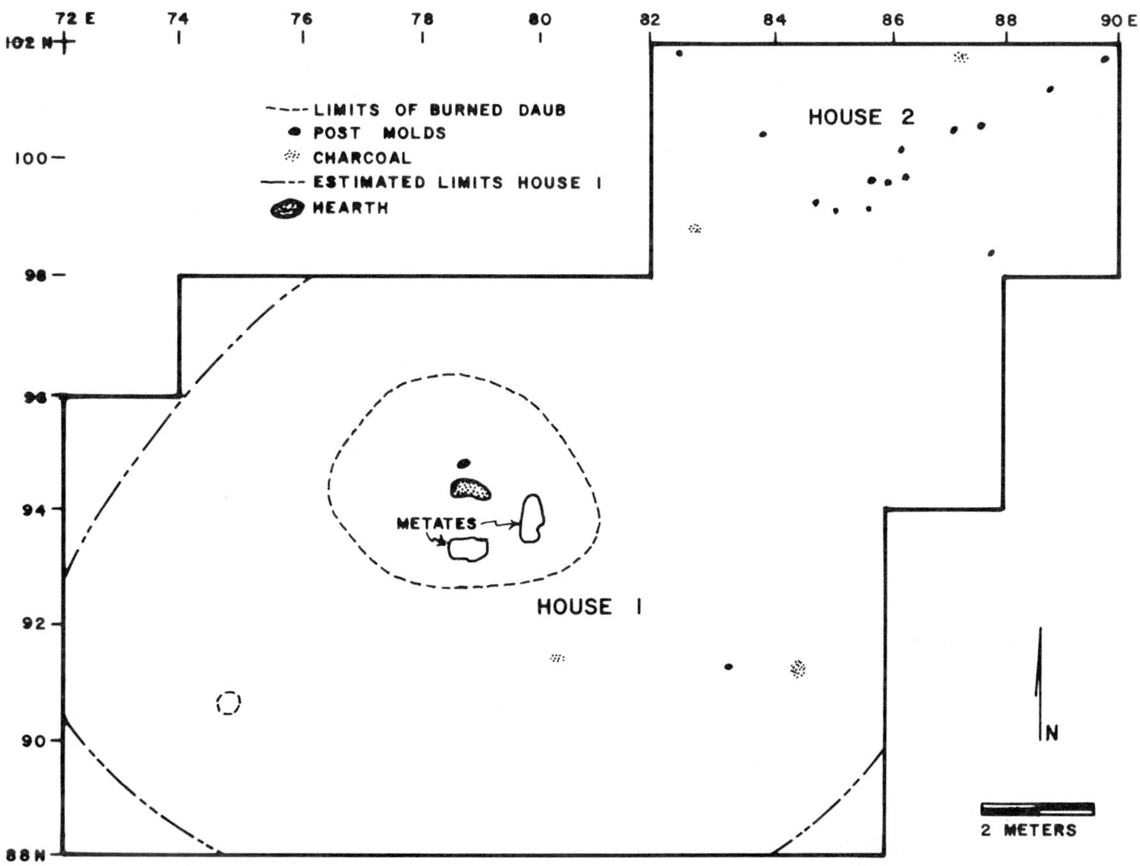

Fig. 5.2 Block excavation, Two Deer site (14BU55); 1978-1979 Investigations.

Table 5.1 Identified genera of charred flora, Two Deer site.[a]

Name[b]	*Count*	*Common Name*
Agastache nepetoides	1	giant hyssop
Amaranthus spp. *	876	pigweed
Ambrosia spp. *	7	ragweed
Arenaria sp.	1	sandwort
Asimina triloba *	1	pawpaw
Celtis sp.	2	hackberry
Chenopodium hybridum *	171	maple-leaved goosefoot
Chenopodium sp. *	2409	goosefoot
Crataequs viridis *	1	red haw
Crotalaria sp.	1	rattle box
Dianthus armeria **	13	pink
Eupatorium sp.	2	thoroughwort
Euphorbia spp. *	241	spurge
Galium sp. *	20	bedstraw
gramineae *	139	grasses
Hypericum perforatum **	31	St. John's wort
Ipomoea sp.	25	morning glory
Kochia scoparia **	6	firebush
Lactuca sp.	1	wild lettuce
Oxalis dillenii	1	wood sorrel
Panicum sp.	26	panic grass
Parthenocissus sp.	3	Virginia creeper
Physalis sp. *	1	ground cherry
Phytolacca americana *	19	pokeweed
Polygonum sp. *	4	smartweed
Portulaca spp. *	31	purslane
Prunus virginiana *	1	plum
Rumex sp. *	1	dock
Scirpus sp.	31	bulrush
Smilax sp.	3	greenbrier
Setaria glauca **	13	yellow foxtail
Silene sp.	2	catchfly
Solanum rostratum	1	nightshade
Stellaria sp.	13	chickweed
Symphoricarpos sp. *	8	coralberry
Viburnum sp. *	7	blackhaw
Viola sp. *	16	violet
Vitis sp *	11	grape
Iva annua *	26	marshelder
Helianthus annuus *	195	sunflower
Zea mays *	114	corn
Cucurbita pepo *	31	squash/pumpkin
unidentified	1918	
unknown	576	
Totals		
identified:	4490	
recovered:	6984	

* probable plant foods— see Table 2.7; ** a species naturalized from Europe or Euroasia.

[a] identifications were made by the author; Mr. H. A. Stephans, Professor Emeritus, Emporia State University; and Dr. R. I. Ford, Museum of Anthropology, University of Michigan. See Adair and Brown (1981) and Adair (1982) for a more detailed discussion of the floral remains.

[b] Taxonomy follows Steyermark (1963) or Baker (1969).

species, such as *Asimina triloba* and *Prunus virginiana*, probably also are food remains although they are represented by low frequencies. In addition, the presence of other species such as *Phytolacca spp.* and *Scirpus sp.* may be due to prehistoric collection and utilization although no data exist to document that these seeds represent food resources. The 17 species listed represent the most likely of all identified taxa to be food sources, based on their abundance within the archaeological deposit and their documented use in ethnographic accounts (e.g. Gilmore 1977, Yanovsky 1936).

The amount of matrix floated at Two Deer represents approximately 12 percent of the portion excavated, which is significantly more than usual in most excavation methodologies. Two reasons prompted this strategy. First, as a means of demonstrating concentrations of particular taxa, a pre-determined amount of matrix had to be floated from each unit. The half bushel size was selected since it has been suggested as a standard measure (Asch and Asch 1980). Second, test excavations (Fulmer 1977) had indicated the site to be rich in charcoal which was somewhat unusual for a shallowly buried Early Ceramic occupation. Extensive

Table 5.2. Identified genera of charred nutshell, Two Deer site.

Name	*Count/Weight*	*Common Name*
Juglans nigra	3901/ 75.9 g	black walnut shell
	149/ 8.5 g	black walnut meat
	627/ 8.4 g	black walnut husk
Carya sp.	62/ 4.4 g	hickory shell
Quercus sp.	151/ 8.5 g	oak shell
	2/ .3 g	oak meat
	2/ .2 g	oak cup
unidentified nutshell	576/ 18.9 g	
Totals		
identified: shell	4116/ 88.8 g	
meat	151/ 8.8 g	
husk	627/ 8.4 g	
cup	2. .2 g	
recovered: shell	4692/107.7 g	

flotation was the best way to retrieve and quantify these data. In retrospect, the control samples proved to be extremely important in demonstrating floral utilization and cultivation since anticipated multiple storage pits were not uncovered.

Table 5.3 lists the ubiquity index for the 17 identified food resources. This index is a measure (listed as a percentage) of how frequently a particular species is identified from the total number of flotation samples. It is determined by dividing the amount listed under the Number of Samples column by the number of flotation samples. Samples taken from test units surrounding the block excavation were not included. Two percentages were calculated; one which included all samples from the main excavation and a second which was figured on samples corresponding with the interior of House 1. As the various percentages indicate, none of the 17 taxa were particularly ubiquitous in the flotation samples; *Juglans nigra* and *Chenopodium spp.* were the most common. Although the actual amounts of these plant remains are quite high, the indices suggest that they were concentrated in certain areas within the house and not randomly distributed.

Table 5.3. Ubiquity Index of Charred Taxa from Two Deer.

Taxa	*Identified in Number of Samples*	*Total %*[a]	*House Interior %*[b]
Juglans nigra	119	31	36
Chenopodium spp.	83	12	25
Zea mays	40	10	12
Cucurbita pepo	19	5	6
Helianthus annuus	24	6	7
Iva annua	11	2.8	3.3
Amaranthus spp.	30	8	9
Portulaca sp.	10	2.5	3
Euphorbia spp.	47	12	14
Polygonum sp.	4	1	1.2
Vitis sp.	9	2.3	2.7
Viola sp.	4	1	1.2
Galium sp.	16	4.1	4.8
Ambrosia spp.	6	1.5	1.8
gramineae	17	4.3	5
Carya sp.	9	2.3	2.7
Quercus spp.	27	7	8

[a] represents 390 samples; [b] representes 332 samples

Features at Two Deer consisted of shallow post molds, a centrally located hearth and a shallow, poorly defined storage pit (Fig. 5.2). While the soil from all of these features was floated, the total number of recovered artifacts was relatively insignificant. Table 5.4 lists the identified flora recovered from the features. Except for the quantity of black walnut shells in the hearth fill and a small concentration of *Chenopodium sp.* and *Amaranthus sp.* in Feature 20, features were relatively void of charred plant remains. If only the contents of intramural features had been floated, which is often the standard recovery procedure, the role of cultivation at this site would have been reduced to one fragmented corn kernel. It may be more than a coincidence that cultigens recovered from the other Plains Woodland sites discussed in this study consist of only one or two corn kernel fragments.

Features, especially storage pits and hearths, have often been assumed to represent capsules of information on prehistoric subsistence since both could easily have been connected with food preparation, cooking, and storage activities. Additionally, pit contents are often interpreted as trash which was periodically gathered from the living area and systematically dumped into a former storage facility. There are however two major problems in assuming such features will always contain sufficient information on subsistence in Plains Woodland sites. First, as discussed in the previous chapter, many Woodland period sites are shallowly buried and therefore subject to various disturbances. These disturbances often destroy the upper portions of the occupations and make feature identification difficult. Even when pits are recognized, their shape and size are seldom as clearly defined as large Middle Ceramic bell-shaped storage facilities. The recovery of macrobotanical remains may easily be biased by these situations. Second, features may be the result of prehistoric behaviors other than those related to food resources (O'Malley 1979; Vickery 1976). Hearths could have functioned as a source of heat for warmth or for thermal alteration of chert and storage facilities need not have stored only food. While these features may contain some food remains (for example corn cobs and nut shells as fuel for fires), their recovery and identification may provide more information on refuse practices than on subsistence. More information on subsistence may be gained by floating soil matrix from areas of food processing, consumption, and discard since all of these activities are likely to result in concentrations of floral remains. Until a pattern in these activities can be demonstrated to exist within a particular archaeological site, an *a priori* assumption that subsistence information will be contained only within intra-mural features will result in an insufficient reconstruction of this strategy.

Systematic recovery techniques of macrobotanical remains from the Two Deer site allow for a reconstruction in the distribution of identified taxa in relation to the features. Figures 5.3 through 5.8 demonstrate the arrangement of the four cultigens, *Chenopodium spp.*, and black walnut within the interior of House

Table 5.4. Identified Flora Recovered from Features at Two Deer.

Feature	Contents: charred	Contents: fresh
#1 charcoal concentration		*Chenopodium hybridum 2*
		Amaranthus sp. 2
#4 post mold		*Chenopodium sp. 6*
#6 post mold	*Chenopodium sp. 4*	*Chenopodium sp. 2*
	*Dianthus armeria 1**	*Amaranthus sp. 31*
	Euphorbia sp. 1	*Setaria glauca 3**
#13 limestone concentration		*Chenopodium sp. 2*
# 16 post mold		*Mollugo verticillata 1*
		Amaranthus sp. 1
#17 post mold		*Amaranthus sp. 5*
#19 hearth	*Juglans nigra 274*	*Amaranthus sp. 60*
	Quercus sp. 7	*Chenopodium spp. 40*
	Chenopodium sp. 56	*Silene sp. 1*
	Euphorbia sp. 2	
	Zea mays 1	
	Parthenocissus sp. 3	
	Galium sp. 2	
	Rumex sp. 1	
	*Setaraia glauca 1**	
	Silene sp. 2	
	Panicum sp. 1	
	Phytolacca americana 1	
	gramineae 5	
	*Dianthus armeria 1**	
#20 storage pit	gramineae 11	
	Amaranthus sp. 78	
	Juglans nigra 1	
	unidentified 2	
	Chenopodium sp. 55	

* a species naturalized from Europe or Euroasia, accidentally charred and included in the archaeological sample. For a discussion of contamination of deposits by modern seeds see Keepax (1977).

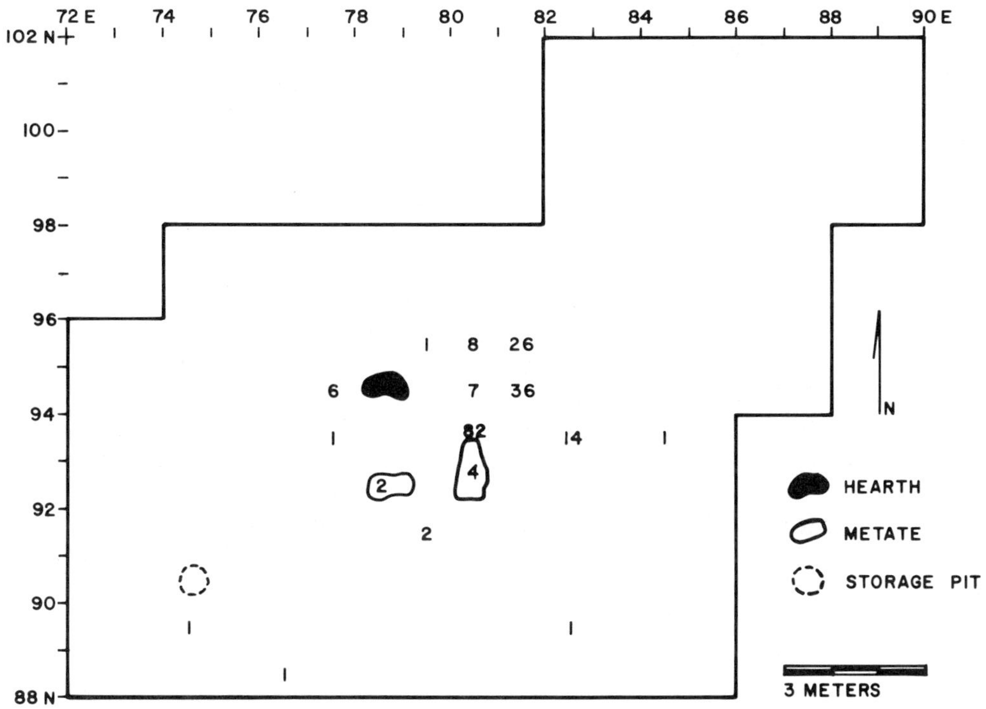

Fig. 5.3 Distribution of *Helianthus annuus* within House 1, Two Deer site.

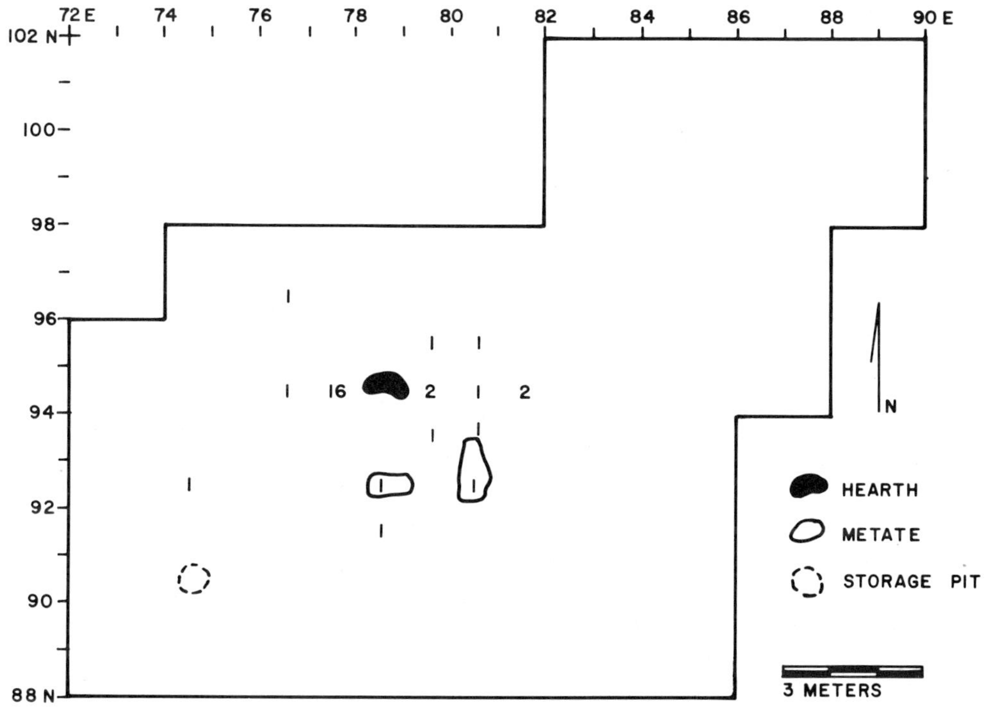

Fig. 5.4 Distribution of *Cucurbita pepo* within House 1, Two Deer site

1. These six taxa were chosen since they represent either extensive utilization or domestication. Actual quantities of charred seeds and weights of nutshells are listed in these figures which clearly illustrate the concentrations of subsistence remains in several areas throughout the structure. There is a definite tendency for charred remains to be associated with the largest metate, although not exclusively. Sunflower achenes and *Chenopodium spp.* seeds in particular (Figs. 5.3 and 5.7) are concentrated in the unit directly to the north of this artifact; a concentration of sunflower achenes was actually observed in this unit during excavation. The curcurbits, maize kernels, and *Iva* achenes are likewise dispersed around the features but are most abundant to the west of the hearth (Figs. 5.4, 5.5, 5.6). This concentration may be the only one resulting directly from the number of samples collected for flotation and may therefore not be a direct reflection of dispersal patterns. Eight flotation samples were taken from this unit, twice as many as from some of the other units. Extra flotation samples however were also taken from other units surrounding the metates and hearth and a similar concentration of the three cultigens is not evident.

Charred *Chenopodium* seeds constitute the most numerous of all identified taxa from Two Deer. As discussed in Chapter 3, recent studies of remains from several eastern sites have demonstrated the prehistoric domestication of this weedy annual (Asch and Asch 1985, Fritz 1984, Smith 1984). Originally thought to involve two species of this plant, the domestication process now focuses on *C. berlandieri* with *C. bushianum*, which was first identified as a domesticate in west central Illinois, relegated to a subspecies or variety (Asch and Asch 1985: 174). Domesticated *Chenopodium* is recognized in Middle Woodland deposits in west central Illinois (Asch and Asch 1982) and northeastern Alabama (Smith 1984) and in later (Mississippian) deposits in Arkansas (Fritz 1984). Interrelated morphological characteristics associated with seed coat thickness and surface patterning clearly establishes *C. berlandieri* as a domestic variety and separates it from its wild relatives.

Despite the large quantity of *Chenopodium* seeds at the Two Deer site, an argument for domestication of this plant cannot be made. Most of the remains are not identified to species level although more than one species is recognized. Several definite concentrations of *Chenopodium spp.* are noted within the excavation, although the seeds are far more random in their distribution than the recognized cultigens (Fig. 5.7). The overall quantity of the seeds however, along with their distribution confined within the limits of House 1, argues for prehistoric collection and utilization rather than to accidental inclusion and charring in the midden.

The distribution of *Juglans nigra* shell fragments is also primarily confined to the center of the house and surrounds the hearth and metates (Fig. 5.8). One of the heaviest concentrations is actually located within the hearth fill, suggesting that this by-product of a dominant food resource may have been used as fuel for the fire.

The function of features appears to be a critical factor in the eventual presence and quantity of botanical remains and their subsequent recovery by flotation. The data from Two Deer suggest that the recovery of charred remains from only the features may lead to an inadequate reconstruction of subsistence since they may not have functioned directly in food preparation, cooking, and storage and therefore unlikely to contain floral remains. Instead of concentrations

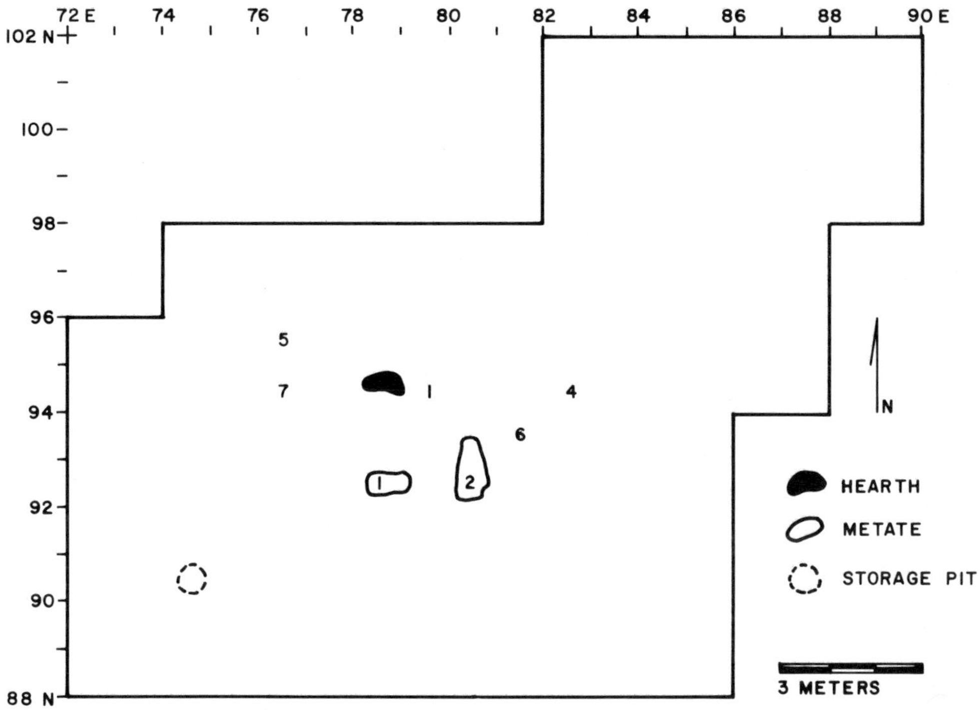

Fig. 5.5 Distribution of *Iva annua* within House 1, Two Deer site.

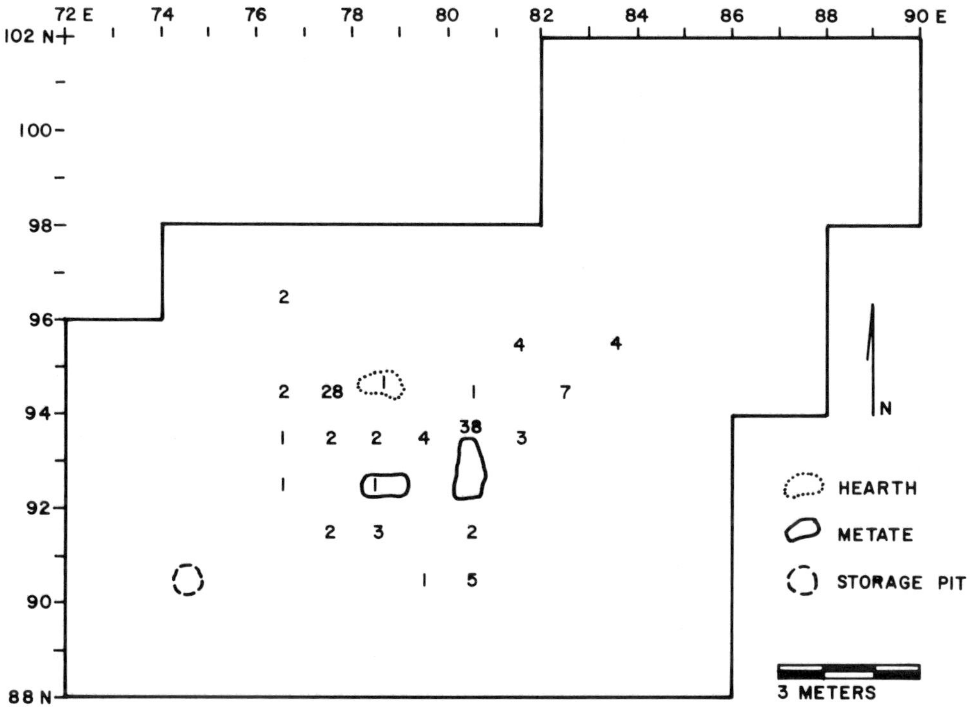

Fig. 5.6 Distribution of *Zea mays* within House 1, Two Deer site.

within features, the macrobotanical remains are present throughout the interior of the house and are most numerous in areas directly adjacent to the grinding slabs. Since features at Two Deer are not numerous however, an extrapolation of these conclusions to Plains Woodland sites in general may be unwarranted. While this may be true for storage pits, the hearth at Two Deer was clearly defined and all contents were floated, suggesting that the hearth may have served multiple purposes and cooking food may have been only one. The style or manner of cooking however would be crucial in the subsequent recovery of charred plant remains.

Braun (1981) has recently suggested that changes in ceramic design may have been correlated with changes in cooking styles from parching or popping in a fire to boiling or simmering over a fire. Increased wall curvature and decreased wall thickness observed in late Woodland ceramics in Illinois served to increase the pot's resistance to thermal stress which allowed for foods to be boiled and simmered in pots over a fire for long periods of time. This was accomplished by suspending a pot over a fire or placing it directly in a fire. If the technological designs were successful, the pot would not break during the cooking stage and food would therefore not likely fall into the fire from this aspect of preparing a meal. Keeping in mind that some by-products of food (i.e. corn cobs, nut shells) may be added to a fire as fuel, it is suggested that hearths associated with this kind of food preparation will contain less charred plant remains than those features into which food was directly placed.

The lack of sufficient quantities of flora in the hearth fill from Two Deer may indicate that foods were cooked from a pot suspended over the fire. In support of this suggestion are the remnants of a charred post in proximity to the hearth (Fig. 5.2) which may have functioned as a scaffold for the cooking vessel. In addition, 11 percent of the rims represent globular shaped vessels, a trait corresponding to Braun's (1981) increased wall curvature. How these variables may appear on a larger scale (throughout the study area) cannot be adequately addressed here. There is some evidence however that ceramics were changing in shape, thickness, and surface treatment during the latter part of the Plains Woodland period. Rims indicative of globular vessels are minimally represented at several Grasshopper Falls phase sites (Reynolds 1979) and at several Pomona variant sites, including the Roth site dated to A.D. 720 and A.D. 850 (Brogan 1982). While not quantitatively defined for the entire Plains Woodland period, ceramics do tend towards decreases in wall thickness in the later centuries (Johnson personal communication). A more obvious change and one perhaps related to thickness, is surface treatment. Earlier Plains Woodland ceramics are cordmarked, with the markings indicative of a cord-wrapped stick meticulously applied to the wet clay in even and consistent intervals. Later Plains Woodland ceramics are also cordmarked, although the erratic and uneven placement of the treatment suggests that a cord-wrapped paddle was used in this step. Hurley (1979) suggests that the use of a paddle and anvil in the manufacture of ceramics serves to compact the clay before firing, thus decreasing its thickness. Anvil marks on the interior of several late Plains Woodland ceramic assemblages are in agreement with this mode of manufacture. It is premature at this time however to discuss changes in Woodland ceramics or to draw an association between thin ceramics and the manner of cooking in the Central Plains. Braun's

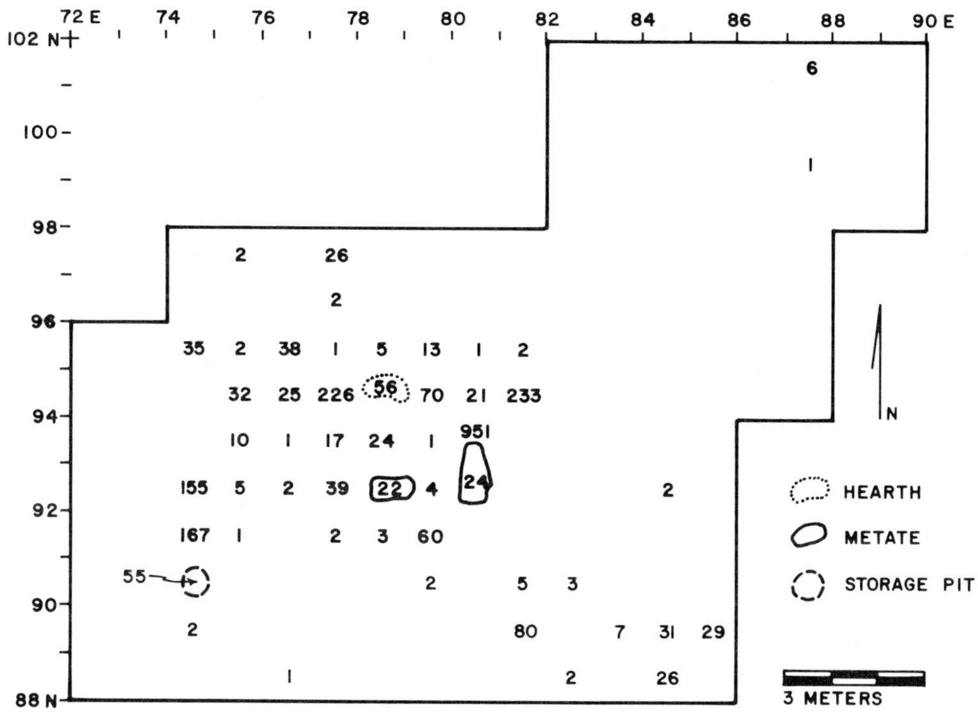

Fig. 5.7 Distribution of *Chenopodium spp.* within House 1, Two Deer site.

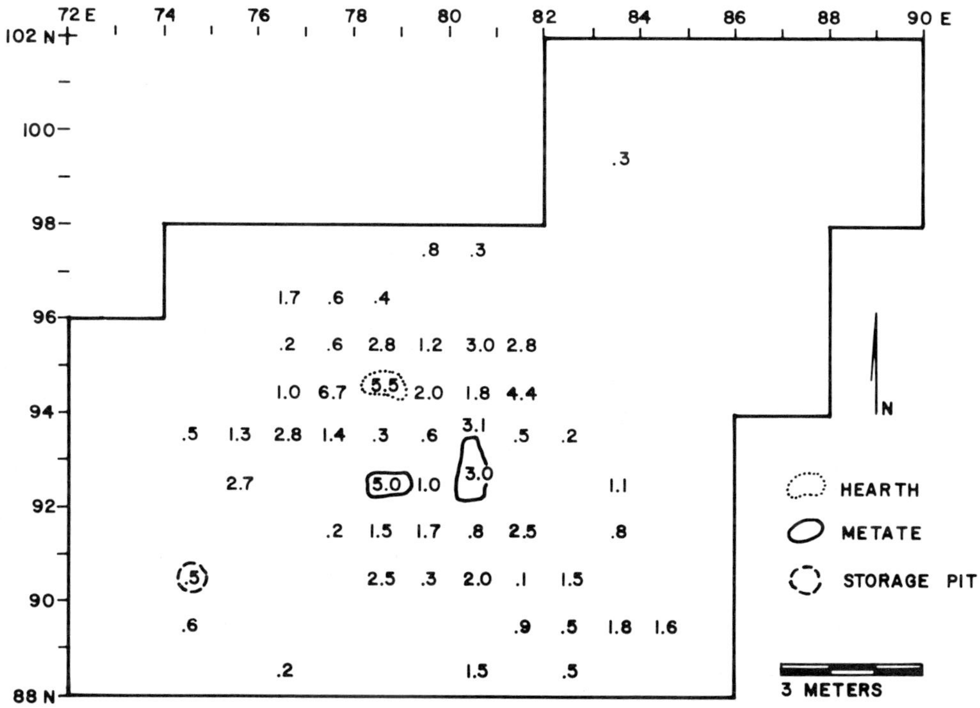

Fig. 5.8 Distribution of *Juglans nigra* within House 1, Two Deer site.

study (1981) has suggested some intriguing paths for future analyses with a recent contribution (Johnson 1984) providing a basis for the Central Plains.

Major changes in lithic and ceramic styles, length of sedentism, social organization, and subsistence may have occurred during the latter part of the Plains Woodland period. However, our understanding of these developments throughout the Woodland period is not clear and can best be addressed through major research. Data presented in this chapter suggest that the sampling procedure used for the recovery of charred plant remains constitutes a significant bias in the recognition of agriculture during the late Woodland. At Two Deer, the distribution of identified cultigens demonstrates concentrations throughout the house interior but not within defined intra-mural features. This is especially evident within the hearth. The lack of charred plant remains in this feature (and to a lesser extent within the storage pit) represents a change from preceding cultural periods. At the Archaic occupations of Coffey (Schmits 1978) and Nebo Hill (Root 1979) as well as the Early Woodland occupation at Traff (Adair 1980), the charred remains were abundantly located within the hearth fills. The difference in location of plant remains between late Plains Woodland and earlier cultures may center around the manner of cooking (boiling in ceramics vs directly placing food in fires) and the mode of discard (the tendency for corn cobs and nutshells to be used as fuels). These changes may be directly observed only by systematic flotation techniques of both feature and midden areas.

CHAPTER 6

Summary And Conclusions

In 1966, a major conference on hunter-gatherers emphasized the point that human populations had, for 99 percent of their existence, secured their food by hunting animals and gathering local plants (Lee and DeVore 1968). In addition, this subsistence strategy was regarded as the most successful and persistent adaptation ever developed. The conference and subsequent articles also stressed the conservative nature of humans; that is, the unlikelihood of change from a successful, healthy, nutritionally balanced, and non-stressful subsistence strategy. Studies on agricultural groups, extant and prehistoric, have supported these descriptions by noting increased evidence of disease, infant mortality, malnutrition, and general biological stress among sedentary agriculturalists over the more mobile hunter-gatherers. These studies have prompted many researchers to theorize that the etiology of agriculture was surrounded by a force or stress situation, actually causing human populations to change against their preference.

A growing body of research and literature however suggests instead that such a change was gradual, uneventful, and probably initially occurred with little recognition by the prehistoric populations. The cultivation and eventual domestication of plants was an evolutionary process which developed from a successful symbiotic relationship between human populations and plants. Rindos (1980, 1984) has grouped this relationship into two systems; domesticatory and agricultural with a primary difference between them being the contribution of the behavior associated with each system to the overall subsistence of the society. Domesticatory systems, which are not restricted to humans alone, but may describe several plant-animal relationships, may be conceptualized for human populations as gardening, tending, or protecting, in which certain cultigens are used for food resources but do not constitute a major subsistence item. Agricultural systems, on the other hand, involve a higher importance level for the plants in the subsistence strategy as well as greater involvement by humans in their growth, maintenance, and productivity. This system may be pictured as a field where the plants contribute to a major portion of the diet. From a biological perspective, the genetic expression in plants within a domesticatory system may be quite variable, a situation which tends to lessen predictability about yields and prohibits the crop from becoming a dependable resource. Greater uniformity in phenotypic and genotypic expression is generally associated with well tended and carefully maintained crops and indicates a product grown for high productivity and dependability, as exemplified for agricultural systems. Agricultural systems are not however, an inevitable outgrowth of domesticatory systems, although they do evolve from them. In prehistory, the change from one system to the other was certainly gradual and may have coincided with other developments or changes, some of which may have been contributing elements in the emergence of agricultural systems. Variability in geographical regions and cultural groups may suggest that different factors should perhaps be ranked in their importance as contributors in the development of agriculture.

This study has focused on the development of agriculture in the Central Plains and has suggested potential factors in this process. Although cultivation appeared relatively late in the study area (as compared with the Southwest and Eastern Woodlands), the processes involved in this change may have been similar to those noted from other areas. A regional methodological approach followed in this research may be the most appropriate for studying processes of cultural change and evolution, particularly given the variability in the recovery of subsistence data and in the amount of excavation and research conducted in more localized areas in the Central Plains. The development of agriculture has been argued from an evolutionary perspective in which the conservative nature of human populations is upheld. It has also taken into account several pertinent factors recognized in the Central Plains. Included are the ecological parameters; the relationship between population, sedentism, and available resources; the genetic "fitness" of domesticated plants; and the population's awareness of the potential subsistence productivity offered by plants.

Six cultigens have been recognized in Central Plains archaeological sites dating to pre-contact periods. Included are the tropical cultigens corn (*Zea mays*), squash (*Cucurbita pepo*), gourd (*Lagenaria siceraria*), and beans (*Phaseolus vulgaris*), and the native cultigens sunflower (*Helianthus annuus*), and marshelder (*Iva annua*). Ongoing research may identify the presence of other cultigens in archaeological deposits within the study area. Attention is presently focused on identifying cultigen *Chenopodium* in Archaic, Early Woodland and Middle Woodland Hopewell deposits and cultivated barley (*Hordeum pusillum*) in Late Woodland contexts. The appearance of the six known cultigens in archaeological deposits throughout the study area can be divided into temporal and spatial groupings. The first recognition of domesticated plants comes from the eastern portion of the Plains, mainly around the Kansas City area and within the Osage Plains. Influence from Eastern Middle Woodland populations was suggested to be a factor in the initial appearance of cultigens, especially maize, squash and *Iva*. A temporal hiatus exists in the occurrence of each of these plants during the Early Ceramic period, a factor strongly suggested to be related to recovery techniques. By A.D. 900-1000, these three cultigens, along with gourd, sunflower, and beans are recorded from a variety of sites throughout the Central Plains and corresponds to the recognition of developed agricultural systems.

Beginning around A.D. 250 the domestication of maize, squash, and *Iva* is recongized from the Middle Woodland Trowbridge site. Maize and *Iva* remains have also been recovered from other Kansas City Hopewell sites (Renner and Yeo) as well as a central Missouri Hopewell site (Fisher-Gabert) and sites from complexes culturally affiliated with the Middle Woodland period (Schultz and Cuesta). These domesticated plants were initially added to the resource inventory because they possessed qualities in nutrition, rate of maturity, and manner of collecting and processing that were similar to foods already collected and consumed by prehistoric populations. These newer foods were accepted within a conservative realm; they were added not as a change to the subsistence economy but as a selection on the existing subsistence base. They provided variety, added economic security, and nutritional benefits but were not staples in the subsistence base and did not initially require changes in technological, social, or settlement levels. The selection for hunting at this time as a means of acquiring necessary energy may have been more pronounced in the Kansas City region since the area does not attract rich and reliable aquatic resources. Population growth and density within the Kansas City area noted in the archaeological record after A.D. 250, may have caused an imbalance between available resources and group size. It was suggested that this problem was temporarily and periodically alleviated by either adjustments in group size and/or settlement location. Both alternatives would have retained the existing settlement-subsistence strategy. An increase in the availability and consumption of domesticated plants was not a high ranking alternative since ecological, demographic, technological, and social conditions favored the continuation of the existing system.

As human groups gradually expanded throughout the study area and the settlement became more permanent, populations in the Central Plains, particularly those west of the Dissected Till and Osage Plains, became more dependent upon river valley resources due to the high biomass levels in these areas. At some point in this particular situation (around A.D. 800-1000), the level of importance of cultigens was increased. Data suggest that factors involved in this selection centered around population density, available resources, and the existing environmental parameters. Agricultural systems are first recognized in three areas in the Central Plains; northcentral Kansas (Solomon River phase); southcentral Kansas (Bemis Creek phase); and the Kansas City area (Steed-Kisker phase). The first two areas are applicable to the following discussion, the third will be summarized separately.

In the study area, west of approximately the 100th meridian, rainfall is too variable and inadequate to support non-irrigation agriculture in zones other than the fertile alluvial valleys. In prehistory, cultivated plants could have been maintained only by inhabiting these locations. Prehistoric populations, who had gradually settled into a sedentary lifestyle during the Woodland period were also dependent upon these areas for annually renewable resources. The extent to which annual or periodic hunting parties exploited upland game resources as part of the subsistence strategy is unknown, although the presence of bison and elk remains in habitation deposits suggests that this strategy was an important element in the economy. The sedentary hamlets or villages in the alluvial valleys were however dependent upon locally available and reliable resources for their survival. The co-occurrence of major group habitations and desired critical resources in the valley areas may have caused an economic and a demographic pressure situation when population densities threatened the ability of the local habitat to maintain the commodities in steady supply. This situation created a subsistence instability between plants and human populations which may have been eliminated by several alternatives. The increase in the availability of cultigens was argued in this study as a conservative choice made by prehistoric populations who were economically tied to the river valleys but who needed a greater quantity of food than the narrow gallery forests could provide on a yearly basis. An increase in the availability of these resources meant that the environmental constraints inhibiting the growth of the cultigens were removed, accelerated by clearing of the native vegetation for the planting of crops. Human's long term symbiotic relationship with plants and particularly with domesticates, and their awareness of the potential yield of these foods, were important cognitive factors that guided the development. This may have coincided with the emergence or introduction of certain positive genetic traits in domesticated plants, most noticeable in maize. Climatic stability, measured by the frequency and intensity of thunderstorms, may have provided additional support in the shift toward increased use of cultigens. Habitats were altered so that crops could prosper. These were the same habitats that had provided the prehistoric inhabitants with the annually renewable resources that once constituted significant portions of the food base. Such an alternative to increase the food supply at the expense of these plants would not have been taken if prehistoric populations had not been aware of the reliability and dependability of agricultural crops. Given the sedentary lifestyle of the dense populations, this alternative was the most conservative one since it allowed for a continuation of group autonomy and a sedentary existence. Other alternatives to increase the food supply, such as trade or an increase in hunting, would have resulted in either greater dependency on surrounding populations or an increase in mobility, both which would have been radical changes.

Botanical, ceramic, and skeletal data combine to suggest that the agricultural system in the Kansas City area may have been partially related to a migration of people or a diffusion of ideas into the area. This suggestion is somewhat tentative given the unknown status of agriculture during the Late Woodland period in the region. Data clearly indicate however that the Steed-Kisker phase included characteristics similar to those of Middle Mississippian populations to the East.

Agricultural systems undoubtedly had profound effects on other cultural aspects, as evinced by changes in the structural design of permanent houses, lithic and ceramic technology and design, and social organization. The extent to which these correlate with the development of agriculture in the Central Plains is however largely unknown and awaits extensive research beyond the scope of this study.

The basis for this study was the organization of archaeobotanical remains from sites throughout the Central Plains. It is appropriate therefore that this manuscript end with a discussion of these data. Several problems plague our understanding of agriculture in the study area; unsorted flotation matrices, unidentified cultigens, and inadequate recovery techniques are major issues. Additionally, and especially for the Archaic and Early Woodland periods, site excavation has been minimal, often being dictated by the amount and location of cultural resource management projects. Perhaps the most serious problem however, is insufficient sampling of habitation areas due to preconceived ideas on the location of food discard. Features may be one of the best archaeological manifestations for the localization of charred plant remains in pre-Woodland sites. Likewise, large bell shaped storage pits and centrally located hearths in Middle Ceramic earthlodges often provide sufficient amounts of floral remains. These standards may not be true for Woodland deposits however. This study has emphasized that the agricultural systems recognized at A.D. 1000 were the result of long term developments that occurred throughout the Woodland period and especially evident during the latter part of this cultural tradition. Charred plant remains recovered from Woodland sites only weakly support this argument, a factor believed to be directly related to inadequate sampling for the recovery of macrobotanical data. Features such as storage pits and hearths may have served a variety of functions during the Woodland period that may not have always been related to subsistence. Consequently, features may not always contain sufficient data for the reconstruction of this economic practice. In addition, possible changes in ceramic design may have been related to changes in the manner of cooking, which in westcentral Illinois, has been suggested to correlate with an increase in the consumption of starchy foods, especially maize (Braun 1981). Such a change in the Central Plains may be directly related to the amount and variety of charred plant remains recovered from intra-mural hearths. Until hypotheses concerning the function of features in different cultural periods as well as the spatial distribution of food preparation and processing activities are formalized into methodological excavation strategies, the importance of agriculture in late Plains Woodland period sites will remain unclear. The spatial arrangement of botanical data at Two Deer demonstrates that features should not be the only focus for flotation sampling within a habitation structure.

In summary, a regional perspective on Central Plains subsistence has suggested several causes related to the development of agriculture. These have been presented as having profound effects on prehistoric populations, although the various choices made which resulted in both the acceptance of domesticates and in the reliance on them as staple resources, were viewed as the most conservative. Therefore, it is argued that the basic conservative nature of human populations in retaining adaptations which are most successful and persistent was very much a part of the shift from hunting-gathering to agriculture. Specific research goals designed for localized areas or drainages in the Central Plains will certainly refine our understanding of this major development in Plains prehistory; the foundation of this development has now been presented.

Literature Cited

Adair, Mary J.

1977 *Subsistence Exploitation at the Young Site: A Predictive Model for Kansas City Hopewell.* Unpublished Master's Thesis. Department of Anthropology, University of Kansas, Lawrence.

1980 Analysis of Carbonized Floral Remains. Appendix B in *Archaeological Investigations in the Proposed Blue Springs Lake Area, Jackson County, Missouri: The Early Woodland Period.* Submitted to Burns and McDonnell Engineers, Kansas City.

1982 Spatial Patterning at the Two Deer Site (14BU55): A Preliminary Report. In *Archaeological Investigations at El Dorado Lake, Butler County, Kansas (Phase III)*, assembled by Paul E. Brockington, Jr., pp. 305-355. Project Report Series No. 51, Museum of Anthropology, University of Kansas, Lawrence.

1983 A New Archaeological Complex in South-Central Kansas: The Bemis Creek Phase. Paper presented at the 5th Annual Flint Hills/Osage Hills Conference, Topeka.

1984a *Prehistoric Cultivation in the Central Plains: Its Development and Importance.* Ph.D. Dissertation, University of Kansas, University Microfilms, Ann Arbor.

1984b A Quantification of Macrobotanical Remains from Middle Ceramic Sites: New Data from the Witt Site (14GE600). Paper presented at the 42nd Annual Plains Conference, Lincoln.

1985 The Distribution and Importance of *Iva annua* in Central Plains Cultivation. Paper presented of the 7th Annual Flint Hills Conference, Norman, Oklahoma.

Adair, Mary J. and Kenneth L. Brown (eds.)

1986 *Prehistoric and Historic Cultural Resources of Selected Sites at Harlan County Lake, Harlan County Nebraska.* Submitted to the U.S. Army Corps of Engineers, Kansas City District. Kaw Valley Engineering and Development Inc., Junction City, KS.

Adair, Mary J., and Marie E. Brown

1981 The Two Deer Site (14BU55): A Plains Woodland-Plains Village Transition. In *Prehistory and History of the El Dorado Lake Area, Kansas (Phase II)*, edited by Mary J. Adair, pp. 237-351. Museum of Anthropology, University of Kansas, Project Report Series, No. 47, Lawrence.

Anderson, Adrian D.

1961 The Glenwood Sequence: A Local Sequence for a Series of Archaeological Manifestations in Mills County, Iowa. *Journal of the Iowa Archaeological Society* 10: 1-101.

Anderson, Edgar

1956 Man as a Maker of New Plants and Plant Communities. In *Man's Role in Changing the Face of the Earth*, edited by William L. Thomas, Jr., pp. 763-777. University of Chicago Press, Chicago.

Angel, J. Lawrence

1975 Paleoecology, Paleodemography and Health. In *Population, Ecology, and Social Evolution*, edited by Steven Polgar, pp. 167-190. Mouton Publishers, Paris.

Artz, Joe Alan

1981 Test Excavations at El Dorado Lake, 1978. In *Prehistory and History of the El Dorado Lake Area, Kansas* (Phase II), edited by M.J. Adair, pp. 54-168. University of Kansas, Museum of Anthropology, Project Report Series No. 47. Lawrence.

1982 Phase III Investigations at 14BU4. *Archaeological Investigation at El Dorado Lake, Butler County, Kansas (Phase III)*, assembled by P.E. Brockington Jr. pp. 34-91. University of Kansas, Museum of Anthropology, Project Report Series, No. 51 Lawrence.

1983 *The Soils and Geomorphology of the East Branch Walnut Valley: Contexts of Human Adaptation in the Kansas Flint Hills.* Unpublished Master's Thesis, Department of Anthropology, University of Kansas, Lawrence.

Asch, David L., Kenneth B. Farnsworth, and Nancy B. Asch

1979 Woodland Subsistence and Settlement in West Central Illinois. In *Hopewell Archaeology: The Chillicothe Conference*, edited by David S. Brose and N'omi Greber, pp. 80-85. Kent State University Press, Kent.

Asch, David L., and Nancy B. Asch

1978 The Economic Potential of *Iva annua* and its Prehistoric Importance in the Lower Illinois Valley. In *The Nature and Status of Ethnobotany*, edited by Richard I. Ford, pp. 301-341. Anthropological Papers, No. 67, Museum of Anthropology. University of Michigan, Ann Arbor.

1982 A Chronology for the Development of Prehistoric Horticulture in West-Central Illinois. Paper presented at the 47th Annual Meeting of the Society for American Archaeology, Minneapolis.

1985 Prehistoric Plant Cultivation in West-central Illinois. In *Prehistoric Food Production in North America*, edited by Richard I. Ford, pp. 149-204. Museum of Anthropology, University of Michigan, Anthropological Papers, No. 75. Ann Arbor, Michigan.

Asch, Nancy B. and David L. Asch

1980 The Dickson Camp and Pond Sites: Middle Woodland Archaeobotany in Illinois. In *Dickson Camp and Pond: Two Early Havana Tradition Sites in the Central Illinois Valley*, edited by Anne-Marie Cantwell, pp.

152-160, Illinois State Museum, Reports of Investigation, No. 36. Springfield.

1981 Archeobotany of Newbridge, Carlin, and Weitzer Sites—The White Hall Components. Appendix B In *Faunal Exploitation and Resource Selection*, by Bonnie Whatley Styles. Northwestern University Archaeological Program, Evanston.

Asch, Nancy B., Richard I. Ford, and David L. Asch

1972 *Paleoethnobotany of the Koster site: The Archaic Horizons*. Illinois Valley Archaeological Program Research Papers and Illinois State Museum, Reports of Investigation, No. 24. Springfield.

Baker, H.G.

1972 Human Influences on Plant Evolution. *Economic Botany* 26 (1):32-43.

Baker, Scott J., William L. Tibesar and Ross G. Hilma

1985 *Human Skeletal material from 23JA277*. Blue Springs Lake project, Jackson County, Missouri. Report submitted to the U.S. Army Corps of Engineers, Kansas City District.

Barclay, A.S., and F.R. Earle

1974 Chemical Analyses of Seeds IV. Oil and Protein Content of 1253 Species. *Economic Botany* 28 (2):178-236.

Bare, Janet E.

1979 *Wildflowers and Weeds of Kansas*. The Regents Press of Kansas, Lawrence.

Barker, William T.

1969 The Flora of the Kansas Flint Hills. The *University of Kansas Science Bulletin* 48(14):525-584.

Barnes, Ethne J.

1977 *The Calovich Burials (14WY7): The Skeletal Analysis of a Plains Mississippian Population*. Unpublished Master's Thesis, Department of Anthropology, Wichita State University, Wichita.

Beadle, George W.

1980 The Ancestry of Corn. *Scientific American* 242(1):112-119.

Bell, Patricia

1976 *Spatial and Temporal Variability within the Trowbridge Site, a Kansas City Hopewell Village*. Unpublished Master's Thesis, Department of Anthropology, University of Kansas, Lawrence.

Bender, Arnold E.

1978 *Food Processing and Nutrition*. Academic Press, New York.

Benn, David W. (editor)

1981 *Archaeological Investigations at the Rainbow Site, Plymouth County, Iowa*. Report submitted to Interagency Archeological Services, National Park Service, Denver.

Berry, Michael S.

1985 The Age of Maize in the Greater Southwest: A Critical Review. In *Prehistoric Food Production in North America* edited by R.I. Ford, pp. 279-308. Anthropological Papers, Museum of Anthropology, University of Michigan, No. 75, Ann Arbor.

Binford, Lewis R.

1964 A Consideration of Archaeological Research Design. *American Antiquity* 29:425- 441.

1968a Archaeological Perspectives. In *New Perspectives in Archaeology*, edited by Sally R. Binford and Lewis R. Binford, pp. 5-32. Aldine, Chicago.

1968b Post-Pleistocene Adaptations. In *New Perspectives in Archaeology*, edited by Sally R. Binford and Lewis R. Binford, pp. 313-341. Aldine, Chicago.

1983 *In Pursuit of the Past: Decoding the Archaeological Record*. Thames and Hudson, New York.

Binford, Lewis R., and W.J. Chasko, Jr.

1976 Nunamiut Demographic History: A Provocative Case. In *Demographic Anthropology: Quantitative Approaches*, edited by Ezra B.W. Zubrow, pp. 63-144. University of New Mexico Press, Albuquerque.

Blake, Leonard W.

1974 Personal Communication to Patricia J. O'Brien, Kansas State University, Letter dated 4/1/1974.

1980 Personal Communication to Patricia J. O'Brien, Kansas State University, Letter dated 9/12/1980.

1981 Early Acceptance of Watermelon by Indians of the United States. *Journal of Ethnobiology* 1(2):193-199.

Blake, Leonard W., and Hugh C. Cutler

1982 Plant Remains from the King Hill Site (23BN1) and Comparisons With Those from the Utz Site. *The Missouri Archaeologist* 43:86-110.

Blake, S.F.

1939 A New Variety of *Iva ciliata* from Indian Rock Shelters in the South-Central United States. *Rhodora* 41:81-86.

Blakeslee, D.J., and W.W. Caldwell

1979 *The Nebraska Phase: An Appraisal*. J & L Reprint Co. Lincoln.

Blakeslee, D.J., and A.H. Rohn

1982 *Man and Environment in Northeastern Kansas: The Hillsdale Lake Project* Vol. 2. Wichita State University, Department of Anthropology. Report prepared for the U.S. Army Corps of Engineers, Kansas City District.

Bohrer, Vorsila L.

1962 *Ethnobotanical Materials from Tonto National Monument*. Southwestern Monuments Association Technical Series 2:75-114. Coolidge, Arizona.

BOSERUP, ESTER
1965 *The Conditions of Agricultural Growth*. Aldine, Chicago.

BRAIDWOOD, ROBERT J.
1960 The Agricultural Revolution. *Scientific American* 203:130-148.

BRAUN, DAVID P.
1977 *Middle Woodland-(Early) Late Woodland Social Change in the Prehistoric Central Midwestern U.S.* Ph.D. Dissertation, University of Michigan, University Microfilms, Ann Arbor.
1981 Pots as Tools. In *The Hammer Theory of Archaeological Research*, edited by A. Keene and J. Moore, pp. 107-134. Academic Press, New York.

BRAUN, DAVID P., AND STEPHAN PLOG
1982 Evolution of "Tribal" Social Networks: Theory and Prehistoric North American Evidence. *American Antiquity* 47:504-525.

BROCKINGTON, PAUL E., JR.
1977 *Culture Drift and Kansas City Hopewell Stone Tool Variability: A Multisite Analysis*. Ph.D. Dissertation, University of Kansas. University Microfilms, Ann Arbor.

BROGAN, WILLIAM T.
1981 *The Cuesta Phase: A Settlement Pattern Study*. Kansas State Historical Society, Anthropological Series, No. 9. Topeka.
1982 *The Roth Site: An Early Pomona Focus Manifestation in Eastern Kansas*. Kansas State Historical Society, Contract Archeology Publication No. 1. Topeka.

BRONSON, BENNETT
1977 The Earliest Farming: Demography as Cause and Consequence. In *Origins of Agriculture*, edited by Charles A. Reed, pp. 23-48. Mouton Publishers, The Hague, Paris.

BROWN, KENNETH L.
1975a Analysis of Faunal Remains from the Fisher-Gabert Site, A Hopewell Affiliated Middle Woodland Village in West-Central Missouri. Ms. on file, Museum of Anthropology, University of Kansas, Lawrence.
1975b Summary Table of Identified Faunal Remains from the Aker Site (23PL43). Ms. on file, Museum of Anthropology, University of Kansas, Lawrence.
1981 Excavations at the Sperry Site, 23JA85. In *Prehistoric Cultural Resources within the Right-of-Way of the Proposed Little Blue River Channel, Jackson County, Missouri*, assembled by K.L. Brown and R.J. Ziegler, pp. 190-283. Department of the Army, Corps of Engineers, Kansas City.
1984 *Pomona: A Plains Village Variant in Eastern Kansas and Western Missouri*. Ph.D. Dissertation, University of Kansas, University Microfilms, Ann Arbor.

BROWN, LIONEL A.
1967 *Pony Creek Archeology*. Publications in Salvage Archeology, No. 5. Smithsonian Institution, River Basin Surveys, Lincoln.

BROWN, MARIE E.
1982a *Cultural Behavior as Reflected in the Vertebrate Faunal Assemblages of Three Smoky Hill Sites*. Unpublished Master's Thesis, Department of Anthropology, University of Kansas, Lawrence.
1982b Site 14BU57. In, *Archaeological Investigations at El Dorado Lake, Butler County, Kansas, Phase (III)*. University of Kansas, Museum of Anthropology, Project Report Series. No. 51. Lawrence.

BROWN, KENNETH L. AND MARIE E. BROWN
1986 The Archaic, Ceramic and Historic Periods. In *Culture History, Kansas Archaeological Preservation Plan*, Vol. II. Draft edited by K.L. Brown and A.H. Simmons. Office of Archaeological Research, Museum of Anthropology and Center for Public Affairs, University of Kansas, Lawrence.

BROWN, KENNETH L. AND ALAN H. SIMMONS (EDS.)
1986 *Kansas Archaeological Preservation Plan, Culture History*. Draft report, Office of Archeological Research, Museum of Anthropology and Center for Public Affairs, University of Kansas, Lawrence.

BROWN, KENNETH L. AND ROBERT J. ZIEGLER
1981 Prehistoric Cultural Resources within the Right-of-Way of the Proposed Little Blue River Channel, Jackson County, Missouri. Department of the Army, Corps of Engineers, Kansas City.

BROWN, W.L., AND E. ANDERSON
1947 The Northern Flint Corns. *Annals of the Missouri Botanical Garden* 34:1-28.

BUIKSTRA, JANE E.
1977 Biocultural Dimensions of Archeological Study: A Regional Perspective. In *Biocultural Adaptation in Prehistoric America*, edited by R.L. Blakely, pp. 67-84. Southern Anthropological Society Proceedings, No. 11. University of Georgia Press, Athens.
1979 Contributions of Physical Anthropologists to the Concept of Hopewell: A Historical Perspective. In *Hopewell Archaeology: The Chillicothe Conference*, edited by David S. Brose and N'omi Greber, pp. 220-233. Kent State University Press, Kent.

CALABRESE, F.A.
1969 *Doniphan Phase Origins: An Hypothesis Resulting from Archaeological Investigations in the Smithville Reservoir Area, Missouri: 1968*. Unpublished Master's Thesis, Department of Anthropology, University of Missouri, Columbia.

CASSIDY, CLAIRE M.
1980 Nutrition and Health in Agriculturalists and Hunter-Gatherers: A Case Study of Two

Prehistoric Populations. In *Nutritional Anthropology*, edited by Norge W. Jerome, Randy F. Kandel, and Gretel H. Pelto, pp. 117-146. Redgrave Publishing Co., Pleasantville.

CHAMBERS, M.E., AND S.K. TOMPKINS
1977 *The Cultural Resources of Clinton Lake, Kansas: An Inventory of Archaeology, History and Architecture*. Report Submitted to the U.S. Army, Corps of Engineers, Kansas City.

CHAPMAN, C.H.
1980 *The Archaeology of Missouri*, II. University of Missouri Press, Columbia.

CHILDE, V. GORDON
1941 *Man Makes Himself*. Watts & Co., London.

CHOMKO, S.A., AND G.W. CRAWFORD
1978 Plant Husbandry in Prehistoric Eastern North America: New Evidence for its Development. *American Antiquity* 43:405-408.

COHEN, MARK N.
1977 *The Food Crisis in Prehistory: Overpopulation and the Origins of Agriculture*. Yale University Press, New Haven.

CONARD, NICHOLAS, DAVID L. ASCH, NANCY B. ASCH, DAVID ELMORE, HARRY GOVE, MEYER RUBIN, JAMES A. BROWN, MICHAEL D. WIANT, KENNETH B. FARNSWORTH, AND THOMAS G. COOK.
1984 Accelerator Radiocarbon Dating of Evidence for Prehistoric Horticulture in Illinois. *Nature* 308(29):443-446.

COWAN, C. WESLEY
1978 The Prehistoric Use and Distribution of Maygrass in Eastern North America: Cultural and Phytogeographical Implications. In *The Nature and Status of Ethnobotany*, edited by Richard I. Ford, pp. 263-288. University of Michigan, Museum of Anthropology, Anthropological Papers 67, Ann Arbor.
1985 Understanding the Evolution of Plant Husbandry in Eastern North America: Lessons from Botany, Ethnography and Archaeology. In *Prehistoric Food Production in North America*, edited by R.I. Ford, pp. 205-244. Museum of Anthropology, University of Michigan, Anthropological Papers No. 75 Ann Arbor.

CUTLER, HUGH C.
1967 Corn and Squash from Six Sites in North and South Dakota. In *An Interpretation of Mandan Cultural History*, by W. Raymond Wood, pp. 177-182. Bulletin 198, Bureau of American Ethnology, Washington, D.C.
1977 Forward. In *Uses of Plants by the Indians of the Missouri River Region*, by Melvin R. Gilmore. Reprinted, University of Nebraska Press, Lincoln. Originally published 1919, 33rd Annual Report, Bureau of American Ethnology, Washington, D.C.

CUTLER, HUGH C., AND L.W. BLAKE
1969 Corn. In *Two House Sites in the Central Plains: An Experiment in Archaeology*, edited by W.R. Wood. Memoir 6, Plains Anthropologist 14(44):Pt. 2:61-62.
1973a Corn from Cahokia Sites. In *Explorations into Cahokia Archaeology*. Bulletin 7, Illinois Archaeological Survey, Urbana.
1973b *Plants from Archaeological Sites East of the Rockies*. Missouri Botanical Garden, St. Louis.
1976 Corn from Snaketown. *The Hohokam* by Emil W. Haury, pp. 365-366. University of Arizona Press, Tucson.
1983 Identified Botanical Remains. Appendix F. In *Walker-Gilmore: A Stratified Woodland Period Occupation in Eastern Nebraska*. Notebook No. 6, Division of Archeological Research, Department of Anthropology, University of Nebraska, Lincoln.

CUTLER, HUGH C., AND T.W. WHITAKER
1961 History and Distribution of the Cultivated Cucurbits in the Americas. *American Antiquity* 26:469-485.

DENIG, EDWIN THOMPSON
1961 *Five Indian Tribes of the Upper Missouri: Sioux, Arickaras, Assiniboines, Crees, Crows*, edited by John C. Ewers. University of Oklahoma press, Norman.

DONHAM, THERESA K.
1982 *The McPherson Site (23JA143) Jackson County, Missouri*. Report of the Phase III Archaeological Investigations submitted to Jackson County Parks and Recreation.

DUMOND, DON E.
1975 The Limitation of Human Population. *Science* 187: 713-721.

EARLE, TIMOTHY K.
1980 A Model of Subsistence Change. In *Modeling Change in Prehistoric Subsistence Economies*, edited by Timothy K. Earle and Andrew L. Christenson, pp. 1-26. Academic Press, New York.

EARLE, F.R., AND QUENTIN JONES
1962 Analyses of Seed Samples from 113 Plant Families. *Economic Botany* 16(4):221-250.

EL NAJJAR, M.Y. AND A.L. ROBERTSON, JR.
1976 Spongy Bones in Prehistoric America. *Science*, 193.

EYMAN, C.E
1966 *The Schultz Focus: A Plains Middle Woodland Burial Complex in Eastern Kansas*. Unpublished Master's Thesis, Department of Archaeology, University of Alberta.

FALK, CARL
1969a Archaeological Salvage in the Kaysinger Bluff Reservoir, Missouri 1966. Report to the National Park Service, Midwest Region, U.S. Department of the Interior, Omaha.
1969b Faunal Remains. In *Two House Sites in the Central Plains: An Experiment in Archaeology* edited by W. Raymond Wood, pp. 44-50. Memoir 6, Plains Anthropologist 14(pt 2).

FLANNERY, KENT V.
1968 Archeological Systems Theory and Early Mesoamerica. In *Anthropological Archeology in the Americas*, edited by Betty J. Meggars, pp. 67-87. Anthropological Society of Washington, Washington, D.C.

FORD, RICHARD I.
1974 Northeastern Archaeology: Past and Future Direction. *Annual Review of Anthropology* 3:385-413.
1979 Gathering and Gardening: Trends and Consequences of Hopewell Subsistence Strategies. *Hopewell Archaeology: The Chillicothe Conference*, edited by D. Brose and N. Greber, pp. 234-238. Kent State University Press, Kent.
1981 Gardening and Farming before A.D. 1000: Patterns of Prehistoric Cultivation North of Mexico. *Journal of Ethnobiology* 1 (1):6-27.
1985 The Processes of Plant Food Production in Prehistoric North America. In *Prehistoric Food Production in North America*, edited by Richard I. Ford pp. 1-18. Museum of Anthropology, University of Michigan, Anthropological Papers. No. 75. Ann Arbor.

FORD, RICHARD I. (EDITOR)
1978 *The Nature and Status of Ethnobotany*. Anthropological Papers No. 67. Museum of Anthropology, University of Michigan, Ann Arbor.

FRAYER, DAVID W. AND LAWRENCE E. BRADLEY
1979 A Late Woodland burial from Platte County, Missouri. *Plains Anthropologist* 23(83):22-27.

FRISCH, ROSE E.
1978 Population, Food Intake and Fertility. *Science*, 19:22-30.

FRISCH, ROSE E., AND J.W. MCARTHUR
1974 Menstrual Cycles: Fatness as a Determinant of Minimum Weight for Height Necessary for the Maintenance or Onset. *Science* 185.

FRITZ, GAYLE J.
1984 Identification of Cultigen Amaranth and Chenopod from Rockshelter Sites in Northwest Arkansas. *American Antiquity* 49:558-572.

FRYE, J.C
1949 Pleistocene Stratigraphic Sequence in Northeastern Kansas. *American Journal of Science* 247(12):883-889.

FRYE, J.C., AND A.B. LEONARD
1952 *Pleistocene Geology of Kansas*. Kansas Geological Survey Bulletin 99.

FRYE, J.C., AND K.L. WALTERS
1950 *Subsurface Reconnaissance of Glacial Deposits in Northeastern Kansas. Kansas Geological Survey Bulletin 86 (6):141-158.*

FULMER, DARREL W.
1976 *Archaeological Excavations within the El Dorado Reservoir Area, Kansas*. Department of the Interior, National Park Service. Interagency Archaeological Services, Denver, University of Kansas Museum of Anthropology, Lawrence.
1977 *Archaeological Investigations within the El Dorado Area, Kansas, 1975*. Department of the Interior, National Park Service, Interagency Archaeological Services, Office of Archaeological and Historical Preservation. Denver.

GALINAT, WALTON C.
1965 The Evolution of Corn and Culture in North America. *Economic Botany* 19:350-357.
1985 Domestication and Diffusion of Maize. In *Prehistoric Food Production in North America*, edited by R.I. Ford, pp. 245-278. Anthropological Papers, Museum of Anthropology, University of Michigan, No. 75, Ann Arbor.

GILBERT, B. MILES
1980 The Plains Setting. In, *Anthropology on the Great Plains*, edited by W. Raymond Wood and Margot Liberty. University of Nebraska Press, Lincoln.

GILDER, ROBERT F.
1926 *The Nebraska Culture Man*. Henry F. Kieser, Omaha.

GILMORE, MELVIN R.
1931 Vegetal Remains of the Ozark Bluff-Dweller Culture. *Papers of the Michigan Academy of Science, Arts, and Letters,* 14:83-102.
1977 *Uses of Plants by the Indians of the Missouri River Region*. Reprinted, University of Nebraska Press, Lincoln. Originally published 1919, Smithsonian Institution, 33rd Annual Report, pp. 42-154, Bureau of American Ethnology, Washignton, D.C.

GRANGE, ROGER T.
1980 *Archaeological Investigations in the Red Willow Reservoir*. Nebraska State Historical Society, Publication in Anthropology, Number 9, Lincoln.

GROSSER, ROGER D.
1970 *The Snyder Site: An Archaic-Woodland Occupation in South-Central Kansas*. Unpublished Master's Thesis, Department of Anthropology, University of Kansas, Lawrence.
1973 A Tentative Cultural Sequence for the Snyder Site, Kansas. *Plains Anthropologist* 18:228-238.
1977 *Late Archaic Subsistence Patterns from the Central Great Plains: A Systemic Model*. Unpublished Ph.D. Dissertation, Department of Anthropology, University of Kansas, Lawrence.

HAAS, DANIEL R.
1983 *Walker-Gilmore: A Stratified Woodland Period Occupation in Eastern Nebraska*. Notebook No. 6, Division of Archeological Research, Department of Anthropology, University of Nebraska, Lincoln.

HEAVIN, CANDICE G.
1970 Two Nebraska Culture Storage Pits from Doniphan County, Kansas. *Plains Anthropologist*, 15 (47):54-62.

HEFFNER, MICHAEL L.
1974 Temporal Variability of Some Kansas City Hopewell Projectile Points. Ms. on file, Museum of Anthropology, University of Kansas, Lawrence.

HEISER, CHARLES B., JR.
1951 The Sunflower among the North American Indians. *Proceedings of the American Philosophical Society* 95:432-448.
1954 Variation and Subspeciation in the Common Sunflower, *Helianthus annuus*. *American Midland Naturaist* 51 (1):287-305.
1976 *Sunflower*. University of Oklahoma Press, Norman.
1978 Taxonomy of *Helianthus* and Origin of Domesticated Sunflower. In *Sunflower Science and Technology*, edited by Jack F. Carter, pp. 31-53. Agronomy, Madison.
1985 Some Botanical Considerations of the Early Domesticated Plants North of Mexico. In *Prehistoric Food Production in North America*, edited by R.I. Ford, pp. 57-72. Anthropological Papers, Museum of Anthropology, University of Michigan, No. 75, Ann Arbor.

HEISER, CHARLES B., R., DALE M. SMITH, SARAH B. CLEVENGER, AND WILLIAM C. MARTIN, JR.
1969 The North American Sunflowers (*Helianthus*). *Memoirs of the Torrey Botanical Club* 22(3), Durham, N.C.

HOTOPP, JOHN
1978 Glenwood: A Contemporary View. In *The Central Plains Tradition: Internal Development and External Relationships*, edited by Donald J. Blakeslee, pp. 109-133, Report 11, Office of the State Archaeologist, University of Iowa, Iowa City.

HOWELL, NANCY
1976 Toward a Uniformitarian Theory of Human Paleodemography. *Journal of Human Evolution* 5:25-40.

HURLEY, WILLIAM M.
1979 *Prehistoric Cordage*. Aldine Manuals on Archeology, 3, Taraxacum.

JACKSON, RAYMOND C.
1960 A Rcvison of the Genus *Iva* L. *University of Kansas Science Bulletin* 41:793-876.

JANTZ, RICHARD L., DOUGLAS W. OWSLEY, AND P. WILLEY
1978 Craniometric Relationships of Central Plains Populations. In *The Central Plains Tradition: Internal Development and External Relationships*, edited by Donald J. Blakeslee, pp. 144-156. Report 11, Office of the State Archaeologist, University of Iowa, Iowa City.

JOCHIM, MICHAEL A.
1981 *Strategies for Survival: Cultural Behavior in an Ecological Context*. Academic Press, New York.

JOHANNESSEN, SISSEL
1981 *Paleoenthobotany*. In Annual Report 1981, FAI-270 Archaeological Mitigation Project, Department of Anthropology, University of Illinois Urbana-Champaign.

JOHANNESSEN, SISSEL, AND LUCY WHALEY
1980 *Annual Report of Paleoethnobotanical Investigations*. In, Annual Report of Investigations 1980, FAI-270 Archaeological Mitigation Project, University of Illinois at Urbana-Champaign.

JOHNSGARD, PAUL .
1979 *Birds of the Great Plains*. University of Nebraska Press, Lincoln.

JOHNSON, ALFRED E
1968 *Archaeological Investigations in the Clinton Reservoir Area, Eastern Kansas*. Report submitted to National Park Service, Midwest Region, Interagency Archaeological Salvage Program, Lincoln.
1973 Archaeological Investigations at the Budenbender Site, Tuttle Creek Reservoir, North-Central Kansas. *Plains Anthropologist*, 18 (62): Pts. 1-2:271-299.
1974 Settlement Pattern Variability in Brush Creek Valley, Platte County, Missouri. *Plains Anthropologist* 19:107-122.
1976 A Model of the Kansas City Hopewell Subsistence-Settlement System. In *Hopewellian Archaeology in the Lower Missouri Valley*, edited by Alfred E. Johnson, pp. 7-15. University of Kansas, Publications in Anthropology, No. 8 Lawrence.
1979 Kansas City Hopewell. In *Hopewell Archaeology: The Chillicothe Conference*, edited by David S. Brose and N'omi Greber, pp. 86-93. Kent State University Press, Kent.
1983 Late Woodland in the Kansas City Locality. *Plains Anthropologist* 28:99-108.
1984 Temporal Relationships of Late (Plains) Woodland Components in Eastern Kansas. *Plains Anthropologist* 29(106):277-288.
In Press Plains Woodland. In *Handbook of North American Indians*, general editor William G. Sturtevant, Smithsonian Institution, Washington, D.C.

JOHNSON, ALFRED E., AND ANN S. JOHNSON
1975 K-means and Temporal Variability in Kansas City Hopewell Ceramics. *American Antiquity* 40:283-295.

JOHNSON, ALFRED E., AND W. RAYMOND WOOD
1980 Prehistoric Studies on the Plains. In, *Anthropology on the Great Plains*, edited by W.R. Wood and M. Liberty. University of Nebraska Press, Lincoln.

JOHNSON, EILEEN M.
1972 *An Analysis and Interpretation of Floral and Faunal Material from a Kansas City Hopewell Site.* Unpublished Master's Thesis, Department of Anthropology, University of Kansas, Lawrence.
1975 *Faunal and Floral Material from a Kansas City Hopewell Site: Analysis and Interpretation.* Occasional Papers of the Museum, Texas Tech University, No. 36. Lubbock.

JOHNSON, WILLIAM C.
1981 Geomorphic and Archaeologic Evidence of Late Holocene Stream System Change in the Central Plains. Paper presented at the Annual Meeting of the Association of American Geographers, Los Angeles.

JONES, VOLNEY H.
1949 Maize from the Davis Site: Its Nature and Interpretation. Appendix in *The George C. Davis Site, Cherokee County, Texas*, by H. Perry Newell and Alex D. Krieger, pp. 241-248. Society for American Archaeology, Memoir 5, 14(2).

KAPLAN, LAWRENCE
1971 Archeology and Domestication in American *Phaseolus* (Beans). In *Prehistoric Agriculture*, edited by Stuart Stuever, pp. 516-533. American Museum Sourcebooks in Anthropology, New York.
1973 Ethnobotanical and Nutritional Factors in the Domestication of American Beans. In, *Man and His Foods*, ed. by C. Earle Smith, Jr., pp. 75-87.

KATZ, SUZANNA R.
1974 *Kansas City Hopewell Activities at the Deister Site.* University of Kansas, Museum of Anthropology, Research Series, No. 1. Lawrence.

KATZ, S.R., M.L. HEDIGER, AND L.A. VALLEROY
1974 Traditional Maize Processing Techniques in the New World. *Science* 184:765-773.

KAY, MARVIN
1975 *Central Missouri Hopewell Subsistence-Settlement System.* Ph.D. Dissertation, University of Colorado. University Microfilms, Ann Arbor.
1979 On the Periphery: Hopewell Settlement of Central Missouri. In *Hopewell Archaeology: The Chillicothe Conference*, edited by David S. Brose and N'omi Greber, pp. 94-99. Kent State University Press, Kent.

KAY, MARVIN, AND ALFRED E. JOHNSON
1977 Havanna Tradition Chronology of Central Missouri. *Mid-Continental Journal of Archeology* 2(2):195-218.

KAY, MARVIN, FRANCIS B. KING, AND CHRISTINE K. ROBINSON
1980 Cucurbits from Phillips Spring: New Evidence and Interpretations. *American Antiquity* 45:806-822.

KEEPAX, C.
1977 Contamination of Archaeological Deposits by Seeds of Modern Origin with Particular Reference to the Uses of Flotation Machines. *Journal of Archaeological Science*, 4(3):221-30.

KEHOE, ALICE B.
1978 Early Civilizations of Asia and Mesoamerica. *Current Anthropology* 19:204-205.

KING, FRANCES B.
1982a Appendix I: Plant Remains from the Yeo Site (23CL199). In The Yeo Site (23CL199); A Kansas City Hopewell Limited Activity Site in Northwestern Missouri, and Some Theories, by Patricia J. O'Brien. *Plains Anthropologist* 27:54-56.
1982b Preliminary Analysis of Botanical Remains from Phillips Spring. In *Holocene Adaptations Within the Lower Pomme de Terre River Valley, Missouri*, edited by Marvin Kay, pp. 701-728. Report submitted to U.S. Department of the Army, Corps of Engineers, Kansas City
1985 Early Cultivated Cucurbits in Eastern North America. In *Prehistoric Food Production in North America*, edited by Richard I. Ford, pp. 73-98. Museum of Anthropology, University of Michigan, Anthropological Papers, No. 75. Ann Arbor, Michigan.

KING, FRANCES B., AND R. BRUCE MCMILLAN
1975 Plant Remains from a Woodland Storage Pit, Boney Springs, Missouri. *Plains Anthropologist* 20 (68):111-116.

KIVETT, MARVIN F.
1949 Archeological Investigations in Medicine Creek Reservoir, Nebraska. *American Antiquity* 14:278-284.
1952 *Woodland Sites in Nebraska.* Nebraska State Historical Society, Publications in Anthropology 1. Lincoln.
1953 *The Woodruff Ossuary, A Prehistoric Burial Site in Phillips County, Kansas.* Bureau of American Ethnology Bulletin 154, River Basin Survey Papers No. 3:103-141.
1962 Logan Creek Complex. Ms. on file, Nebraska State Historical Society, Lincoln.
1970 Early Ceramic Environmental Adaptations. In, *Pleistocene and Recent Environments of the Central Plains*, edited by Wakefield Dort, Jr. and J. Knox Jones, pp. 93-102. University of Kansas Press, Lawrence.

KOLATA, G.B.
1974 !Kung Hunters and Gatherers: Feminism, Diet, and Birth Control. *Science* 185 (1455):932-934.

KOPSICK, PAUL R.
1982 Geology of the Little Blue Valley. In *Little Blue prehistory: Archaeolgical Investigations at Blue Springs and Longview Lakes, Jackson County Missouri.* Edited by Larry J. Schmits. Report

submitted to the U.S. Army Corps of Engineers, Kansas City District.

KRAUSE, RICHARD A.
1969 Correlation of Phases in Central Plains Prehistory. In *Two House Sites in the Central Plains: An Experiment in Archaeology*, edited by W. Raymond Wood. Memoir 6, Plains Anthropologist 14 (Pt. 2):82-96.
1970 Aspects of Adaptation Among Upper Republican Subsistence Cultivators. In, *Pleistocene and Recent Environments of the Central Great Plains*, edited by Wakefield Dort and J. Knox Jones. University Press of Kansas, Lawrence.

KÜCHLER, A.W.
1964 *Potential Natural Vegetation of the Coterminous United States*. American Geographical Society, Special Publications 36:1-38.

LALLO, J.W., G.J. ARMELAGOS, AND R.P. MENSFORTH
1977 The Role of Diet, Disease, and Physiology in the Origin of Porotic Hyperostosis. *Human Biology* 49:471-483.

LARSEN, CLARK S., AND PATRICIA J. O'BRIEN
1973 The Cochran Mound, 23PL86, Platte County, Missouri. *The Missouri Archaeological Society Newsletter*, No. 267. Columbia.

LEAF, GARY R
1979 Test Excavations Conducted at El Dorado Lake, 1977. In *Finding, Managing, and Studying Prehistoric Cultural Resources at El Dorado Lake, Kansas* (Phase I), edited by Gary R. Leaf, pp. 74-187. Research Series No. 2, University of Kansas, Museum of Anthropology, Lawrence.
1981 Woodland Occupation of the Snyder Site. In *Prehistory and History of the El Dorado Lake Area, Kansas (Phase II)*, edited by Mary J. Adair, pp. 169-236. Project Report Series No. 47, University of Kansas, Museum of Anthropology, Lawrence.

LEAF, GARY R., AND MATTHEW J. ROOT
1979 *Test Excavations Conducted at 14BU32*. Report Submitted to U.S. Army Corps of Engineers, Tulsa.

LEE, RICHARD B.
1972 Population Growth and the Beginnings of Sedentary Life Among the !Kung Bushmen. In *Population Growth: Anthropological Implications*, edited by Brian Spooner, pp. 329-342. MIT Press, Cambridge.

LEE, RICHARD, AND IRVEN DEVORE
1968 *Man the Hunter*. Aldine, Chicago.

LEHMER, DONALD J.
1971 *Introduction to Middle Missouri Archeology*. National Park Service, Department of the Interior, Anthropological Papers No. 1, Washington, D.C.

LIPPINCOTT, KERRY
1978 Solomon River Upper Republican Settlement Ecology. In *The Central Plains Tradition: Internal Development and External Relationships*, edited by Donald J. Blakeslee, pp. 81-93. Report 11, Office of the State Archaeologist, University of Iowa, Iowa City.

LOGAN, BRAD
1979 *Cultural Resources of Kansas City International Airport and it's Environs: An Archaeological Reconnaissance*. Research Series No. 3, University of Kansas, Museum of Anthropology, Lawrence.
1981 *An Archaeological Survey of the Stranger Creek Drainage System, Northeast Kansas*. Project Report Series No. 48, University of Kansas, Museum of Anthropology, Lawrence.

LOGAN, BRAD (EDITOR)
1983 *Archaeological Investigations in the Stranger Creek, Buck Creek, and Mud Creek Drainage Systems, Northeast Kansas, Phase II*. Project Report Series No. 52, University of Kansas, Museum of Anthropology, Lawrence.

LONG, AUSTIN, AND JAMES E. MIELKE
1966 Smithsonian Institution Radiocarbon Measurements III. *Radiocarbon*, edited by Edward S. Deevey and Richard Foster Flint 8:413-422.

LUDWICKSON, JOHN
1978 Central Plains Tradition Settlements in the Loup River Basin: The Loup River Phase. In *The Central Plains Tradition: Internal Development and External Relationships*, edited by Donald J. Blakeslee, pp. 94-108. Report 11, Office of the State Archaeologist, University of Iowa, Iowa City.

MANGELSDORF, PAUL C.
1974 *Corn: Its Origin, Evolution, and Improvement*. The Belknap Press of Harvard University Press, Cambridge.
1983 The Mystery of Corn: New Perspectives. *Proceedings of the American Philosophical Society* 127:215-247.

MANGELSDORF, PAUL C., H.W. DICK, AND J. CAMERA-HERNANDEZ
1967 Bat Cave Revisited. *Harvard University Botanical Museum Leaflets* 22:1-31.

MARSHALL, JAMES O.
1972 *The Archaeology of the Elk City Reservoir: A Local Archaeological Sequence in Southeast Kansas*. Kansas State Historical Society, Anthropological Papers, No. 6, Topeka.

MARTIN, TERRENCE L.
1976 Prehistoric Settlement-Subsistence Relationships in the Fishing River Drainage, Western Missouri. *The Missouri Archaeologist* 37:1-91.

MCDONALD, LEONARD
1967 An Archeological Survey Along McDowell Creek, Geary County, Kansas. *Kansas Anthropological Association Newsletter* 13(3):1-10. Topeka.

McHugh, William P.
1982 *Before Smith's Mill; Archaeological and Geological Investigations in the Little Platte River Valley, Western Missouri*. Report submitted to the Department of the Army, Corps of Engineers, Kansas City.

McKern, W.C.
1939 The Midwestern Taxonomic Method as an Aid to Archaeological Culture Study. *American Antiquity* 4:301-313.

Monger, Earl
1970 A Preliminary Report of the Larned Site. *Kansas Anthropological Association Newsletter* 15(8):1-15. Topeka.

Montet-White, Anta
1968 *The Lithic Industries of the Illinois Valley in the Early and Middle Woodland period*. Anthropological Papers No. 35, Museum of Anthropology, University of Michigan, Ann Arbor.

Munsell, Marvin R.
1961 Anthony: A Kansas-Oklahoma Border Site. *Plains Anthropologist* 6 (Pt. 2):112-114.

Nickel, Robert K.
n.d. Botanical Remains from Sites in Mills County, Iowa. Ms. on file, Midwest Archeological Center, Lincoln.
1977 The Study of Archaeologically derived Plant Materials from the Middle Missouri Subarea. *Plains Anthropologist* 22(78-2):53-58. Memoir 13.

Nickels, Martin K
1971 *An Analysis of the Skeletal Material from Sugar Creek Ossuary (23PL58)*. Unpublished Master's Thesis, Department of Anthropology, University of Kansas, Lawrence.

Oberlander, Theodore M. and Robert A. Muller
1982 *Essentials of Physical Geography Today*. Random House, New York.

O'Brien, Patricia J.
1970 Western Middle Mississippian and its Relation to Cahokia and Central Plains Cultures. Paper Presented at the 36th Annual Meeting of the Society for American Archaeology, Mexico City.
1972 A Preliminary Review of Steed-Kisker Culture. Paper Presented at the 30th Annual Plains Conference, Lincoln.
1973 A New Synthesis of Steed-Kisker (Western Middle Mississippian). Paper presented at the 39th Annual Meeting for the Society for American Archaeology, San Francisco.
1974 A Seriation of Steed-Kisker Ceramics. Paper Presented at the 39th Annual Meeting of the Society for American Archaeology, Washington, D.C.
1976 Steed-Kisker and Mississippian Influences on the Central Plains. Paper presented at the 34th Annual Plains Conference, Minneapolis.
1978a Steed-Kisker and Mississippian Influences on the Central Plains. In *The Central Plains Tradition Internal Development and External Relationships*, edited by Donald J. Blakeslee. Office of the State Archaeologist, University of Iowa, Report 11, Iowa City.
1978b Steed-Kisker: A Western Mississippian Settlement System. In *Mississippian Settlement Patterns*, edited by Bruce D. Smith, pp. 1-18. Academic Press, New York.
1982 The Yeo Site (23CL199): A Kansas City Hopewell Limited Activity Site in Northwestern Missouri and Some Theories. *Plains Anthropologist* 27(95):37-56.
1984 *Archeology in Kansas*. University of Kansas, Museum of Natural History, Public Education Series No. 9, Lawrence.

O'Brien, Patricia, J., Clark Larson, John O'Grady, Brian O'Neill, and Ann S. Stirland
1973 The Elliott Site (14GE303): A Preliminary Report. *Plains Anthropologist* 18(59):54-72.

O'Brien, Patricia J., Margaret Caldwell, John Jilka, Lynn Toburen, and Barbara Yeo.
1979 The Ashland Bottoms Site (14RY603): A Kansas City Hopewell Site in North-Central Kansas. *Plains Anthropologist* 23:1-20.

O'Malley, Nancy
1979 *Subsistence Strategies at the Sperry Site, Jackson County, Missouri*. Unpublished Master's Thesis, Department of Anthropology, University of Kansas, Lawrence.

Padgett, W.M. and D.J. Blakeslee
1982 El Dorado Prehistory: Investigations of some Early Campsites in Butler County, Kansas. Ms. Wichita State University, Archaeology Laboratory, Wichita.

Parks, Sharon G
1978 *Test Excavations at 14GE41: A Schultz Focus Habitation Site at Milford Lake, Kansas*. Report prepared for U.S. Army Corps of Engineers, Kansas City.

Parmalee, Paul W., Andreas A Paloumpis, and Nancy Wilson
1972 *Animals Utilized by Woodland Peoples Occupying the Apple Creek Site, Illinois*. Illinois State Museum, Report of Investigations No. 23. Springfield.

Payne, Willard W. and Volney H. Jones
1962 The Taxonomic Status and Archaeological significance of a Giant Ragweed from Prehistoric Bluff Shelters in the Ozark Plateau Region. *Papers of the Michigan Academy of Science, Arts and Letters* 47:147-163.

Peterson, Robert R., Jr.
1982 The Bowlin Bridge site (23JA38). In *Little Blue Prehistory: Archaeological Investigations at Blue Springs and Longview Lakes, Jackson County, Missouri*. Edited by L.J. Schmits. Report submitted the U.S. Army Corps of Engineers, Kansas City District.

PETERSON, ROBERT R., JR., AND LARRY J. SCHMITS
1982 Phase II test excavations in the Blue Springs and Longview Lake areas. In *Little Blue Prehistory: Archaeological Investigations at Blue Springs and Longview Lakes, Jackson County, Missouri*. Edited by Larry J. Schmits. Report submitted to the U.S. Army Corps. of Engineers, Kansas City District.

PICKERSGILL, BARBARA B. AND CHARLES B. HEISER
1977 Origins and Distribution of Plants Domesticated in the New World Tropics. In *Origins of Agriculture*, edited by C. Reed, pp. 803-835. The Hague: Mouton, Paris.

PLOG, FRED T.
1974 *The Study of Prehistoric Change*. Academic Press, New York.

PULLEN, D.L.
1976 Form and Function of a Feature at the Aker Site (23Pl43). Ms. on file, Museum of Anthropology, University of Kansas, Lawrence.

REEDER, ROBERT L.
1977 *Limited testing of the Coffin Site 23JA200*. Submitted to the Missouri State Highway Commission and Federal Highway Administration.
1980 The Sohn Site: A Lowland Nebo Hill Complex Campsite. In *Archaic Prehistory on the Prairie-Plains Border*, edited by Alfred E. Johnson, pp. 55-66. University of Kansas, Publications in Anthropology, No. 12. Lawrence.

REID, KENNETH C.
1976 Prehistoric Trade in the Lower Missouri River Valley: An Analysis of Middle Woodland Bladelets. In *Hopewellian Archaeology in the Lower Missouri Valley*, edited by Alfred E. Johnson, pp. 66-99. University of Kansas, Publications in Anthropology, No. 8. Lawrence.
1980a *Nebo Hill: Archaic Political Economy in the Riverine Midwest*. Ph.D. Dissertation, Department of Anthropology, University of Kansas. University Microfilms, Ann Arbor.
1980b The Achievement of Sedentism in the Kansas City Region. In *Archaic Prehistory on the Prairie-Plains Border*, edited by Alfred E. Johnson, pp. 29-42. University of Kansas, Publications in Anthropology, No. 12. Lawrence.
1983 The Nebo Hill Phase: Late Archaic Prehistory in the Lower Missouri Valley. In *Archaic Hunters and Gatherers in the American Midwest*, edited by James A. Brown, pp. 11-39. Academic Press, New York.

REID, KENNETH C. AND JOE ALAN ARTZ
1984 *Hunters of the Forest Edge: Culture, Time and Process in the Little Chaney Basin*. University of Tulsa Laboratory of Archaeology, Contributions in Archaeology 14 and Oklahoma Archeological Survey. Studies in Oklahoma's Past, No. 13, Norman.

REIDHEAD, VAN A.
1980 The Economics of Subsistence Change: A Test of an Optimization Model. In *Modeling Change in Prehistoric Subsistence Economics*, edited by Timothy K. Earle and Andrew L. Christenson, pp. 141-180. Academic Press, New York.

RENFREW, COLIN
1972 *The Emergence of Civilization: The Cylades and the Aegean in the Third Millennium B.C.* Methuen, London.

REYNOLDS, JOHN D.
1979 *The Grasshopper Falls Phase of the Plains Woodland*. Kansas State Historical Society, Anthropological Series, No. 7. Topeka.
1984 *The Cow-Killer Site: Melvern Lake, Kansas*. Kansas State Historical Society, Anthropological Series, No. 12, Topeka.
1985 Personal communication.

RINDOS, DAVID
1980 Symbiosis, Instability, and the Origins and Spread of Agriculture: A New Model. *Current Anthropology* 21(6):751-772.
1984 *The Origins of Agriculture: An Evolutionary Perspective*. Academic Press, New York.

ROBINSON, SARAH
1976 An Analysis of Charred Seeds from a Middle Woodland Occupation Site in Central Missouri. In *Hopewellian Archaeology in the Lower Missouri River Valley*, edited by Alfred E. Johnson, pp. 100-109. University of Kansas, Publications in Anthropology, No. 8. Lawrence.

ROEDL, LEO J., AND JAMES H. HOWARD
1957 Archaeological Investigations at the Renner Site. *Missouri Archaeologist* 18:52-96.

ROGERS, RICHARD
1984 *Kansas Prehistory: An Alluvial Geomorphological Perspective*, Ph.D. Dissertation, University of Kansas, University Microfilms, Ann Arbor.

ROHN, A.H., AND C.F. WOODMAN
1976 *Cultural Inventory for Hillsdale Lake, Kansas*, Vol. I. Report Submitted to the U.S. Department of the Army, Corps of Engineers, Kansas City.

ROHN, A.H., C.M. STEIN, AND G. GLOVER
1977 *Wolf Creek Archaeology, Coffey County, Kansas* . Report on file, Wichita State University, Archaeology Laboratory, Wichita.

ROOT, MATTHEW J.
1979 The Paleoethnobotany of the Nebo Hill Site. *Plains Anthropologist* 24(85):239-248
1981a *The Milbourn Site: Late Archaic Settlement in the Southern Flint Hills of Kansas*. Unpublished Master's Thesis, Department of Anthropology, University of Kansas, Lawrence.
1981b Paleoethnobotanical Analysis of Woodland Components on 14BU9 and 14BU31. In *Prehistory and History of the El Dorado Lake Area,*

Kansas (Phase II), edited by Mary J. Adair, pp. 357-372. Project Report Series No. 47, Museum of Anthropology, University of Kansas, Lawrence.

ROPER, DONNA C.
1976 A Trend Surface Analysis of Central Plains Radiocarbon Dates. *American Antiquity* 41:181-188.

ROWLISON, DON D.
1980 *The 1978 Archeological Investigations at Big Hill Lake, Kansas*. Report submitted to the U.S. Department of the Army, Corps of Engineers, Tulsa.

SANTLEY, ROBERT S., AND ERIC K. ROSE
1979 Diet, Nutrition, and Population Dynamics in the Basin of Mexico. *World Archaeology* 11(2):185-207.

SAUER, C.O.
1952 *Agricultural Origins and Dispersals*. The American Geographical Society, New York.

SCHMITS, LARRY J.
1976 *The Coffey Site: Environment and Cultural Adaptation at a Prairie Plains Archaic Site*. Unpublished Master's Thesis, Department of Anthropology, University of Kansas, Lawrence.
1978 The Coffey Site: Environment and Cultural Adaptation at a Plains Archaic Site. *Mid-Continental Journal of Archaeology*, Special Paper No. 1.
1980 The Williamson, Salb, and Dead Hickory Sites. In, *Salvage Archeology of the John Redmond Lake, Kansas*. Kansas State Historical Society, Anthropological Series, No. 8. Topeka.
1981 *Archaeological and Geological Investigations at the Coffey Site, Tuttle Creek Lake, Kansas*. Report submitted to the Interagency Archaeological Services Branch, Rocky Mountain Region, National Park Service, Denver.
1982 The May Brook Site, Jackson County, Missouri. *The Missouri Archaeologist* 43:1-67.

SCHMITS, LARRY J. (EDITOR)
1982 *Little Blue Prehistory: Archaeological Investigations at Blue Springs and Longview Lakes, Jackson County, Missouri*. Report submitted to U.S. Department of the Army, Corps of Engineers, Kansas City.

SCHMITS, LARRY J. (EDITOR)
1986 *Prehistory of the Little Blue River Valley, Western Missouri: Archaeological Investigations at Blue Springs Lake*. Submitted to the U.S. Army Corps of Engineers, Kansas City District.

SCHMITS, LARRY J. AND BRUCE C. BAILEY
1986 Prehistoric Chronology and Settlement-Subsistence Patterns in the Little Blue Valley, Western Missouri. In *Prehistory of the Little Blue River Valley, Western Missouri: Archaeological Investigations at Blue Springs Lake*. Edited by Larry J. Schmits. Submitted to the U.S. Army Corps of Engineers, Kansas City District.

SCHMITS, LARRY J. AND CHRISTOPHER A. WRIGHT
1982 The Turner-Casey site (23JA35). In *Little Blue Prehistory: Archaeological Investigations at Blue Springs and Longview Lakes, Jackson County, Missouri*. Edited by L.J. Schmits. Submitted to the U.S. Army Corps of Engineers, Kansas City District.

SCHOEWE, WALTER
1949 The Geography of Kansas. *Transactions of the Kansas Academy of Science* 52(3):261-333.

SCHROEDER, CHARLES P.
1979 *Nebraska Weeds*. Nebraska Department of Agriculture, Lincoln.

SCHULTZ, FLOYD, AND ALBERT C. SPAULDING
1948 A Hopewellian Burial Site in the Lower Republican Valley, Kansas. *American Antiquity* 13:306-313.

SELF, H.
1978 *Environment and Man in Kansas*. The Regents Press of Kansas, Lawrence.

SHIPPEE, J. METT
1970 A Brief Summary of the Prehistoric Peoples of Clay County, Missouri. *Missouri Archaeological Society Newsletter* 243:2-4.
1972 *Archaeological Remains in the Kansas City Area: The Mississippian Occupation*. Missouri Archaeological Society Research Series No. 9, Columbia.

SMITH BRUCE D.
1978 Variation in Mississippian Settlement Patterns. In *Mississippian Settlement Patterns*, edited by Bruce D. Smith, pp. 479-504. Academic Press, New York.
1984 *Chenopodium* as a Prehistoric Domesticate in Eastern North America: Evidence from Russell Cave, Alabama. *Science* 226(4671): 165-167.

STEINACHER, TERRY LEE
1976 *The Smoky Hill Phase and its Role in the Central Plains Tradition*. Unpublished Master's Thesis, University of Nebraska, Lincoln.

STEVENS, WILLIAM C.
1961 *Kansas Wild Flowers*. 2nd ed. University of Kansas Press, Lawrence.

STEWART, T. DALE
1943 Skeletal Remains from Platte and Clay Counties, Missouri. In *Archeological Investigations in Platte and Clay Counties, Missouri*, by Waldo R. Wedel, pp. 245-273. U.S. National Museum Bulletin 183, Smithsonian Institution, Washington, D.C.

STEYERMARK, JULIAN A.
1963 *Flora of Missouri*. The Iowa State University Press, Ames.

STINI, WILLIAM A.
1975 Adaptive Strategies of Human Populations under Nutritional Stress. In *Biosocial Interelations in Population Adaptation*, edited by E.S.

Watts, F.E. Johnston, and G.W. Lasker, pp. 19-42. The Hague, Mouton, Paris.

Streuver, Stuart
1968 Flotation Techniques for the Recovery of Small Scale Archaeological Remains. *American Antiquity* 33(3):353-362.

Streuver, Stuart, and Kent D. Vickery
1973 The Beginnings of Cultivation in the Midwest Riverine Area of the United States. *American Anthropologist* 75:1197-1220.

Styles, Bonnie W.
1981 *Faunal Exploitation and Resource Selection: Early Late Woodland Subsistence in the Lower Illinois Valley*. Northwestern University Archeological Program, Scientific Papers No. 3, Evanston.

Sussman, R.W.
1972 Child Transport, Family Size, and Increase in Human Population During the Neolithic. *Current Anthropology* 13:258-259.

Swanton, John R.
1946 The Indians of the Southeastern United States. *Smithsonian Institution, Bureau of American Ethnology, Bulletin 137*, Washington, D.C.

Tainter, Joseph A.
1975 *The Archaeological Study of Social Change: Woodland Systems in West Central Illinois*. Ph.D. Dissertation, Northwestern University. University Microfilms, Ann Arbor.

Terra, G.J.A.
1954 Mixed Garden Horticulture in Java. *Journal of Tropical Geography* 3:34-43.

Tiffany, Joseph A.
1981 A Compendium of Radiocarbon Dates for Iowa Archaeological Sites. *Plains Anthropologist* 26(91):55-73.

Tjaden, Rex
1974 The Cogan Mounds, 23PL125, Platte County, Missouri. *The Missouri Archaeological Society Newsletter* No. 284. Columbia.

Tomanek, G.W., and G.K. Hulett
1970 Effects of Historical Droughts on Grassland Vegetation in the Central Great Plains. In *Pleistocene and Recent Environments of the Central Great Plains*, edited by Wakefield Dort and J. Knox Jones, pp. 203-210. The University Press of Kansas, Lawrence.

University of Illinois Agricultural Experiment Station
1976 *Weeds of the North Central States*. Circular 718. University of Illinois. Agricultural Experiment Station, Urbana.

Vickery, Kent D.
1976 *An Approach to Inferring Archaeological Site Variability*. Ph.D. Dissertation, Indiana University. University Microfilms, Ann Arbor.

Voigt, Eric
1985 Paleoethnobotanical Analysis of Material from 23JA269. In *Phase I and Phase II Archaeological Investigations of the Vaughn Estess site(23JA269)*, Jackson County, Missouri, by John M. Parisi. Environmental Systems Analysis Inc., Cultural Resource Management Report, Shawnee Mission, KS.

Wainio, Walter W. and Ernest B. Forbes
1941 The Chemical Composition of Forest Fruits and Nuts from Pennsylvania. *Journal of Agricultural Research* 62:627-635.

Watson, Patty Jo
1985 The Impact of Early Horticulture in the Upland Drainages of the Midwest and Midsouth. In *Prehistoric Food Production in North America* edited by R.I. Ford, pp. 99-148. Anthropological Papers, Museum of Anthropology. University of Michigan, No. 75, Ann Arbor.

Watson, Patty Jo (editor)
1974 *Archeology of the Mammoth Cave Area*. Academic Press, New York.

Watt, B.K., and A.L. Merrill
1963 *Composition of Foods*. U.S. Department of Agriculture, Handbook No. 8. Washington, D.C.

Weakly, H.E.
1943 A Tree-Ring Record of Precipitation in Western Nebraska. *Journal of Forestry* 41(11):816-819.

Wedel, Waldo R.
1935 Reports of Fieldwork by the Archeological Survey of the Nebraska State Historical Society, 1934. *Nebraska Historical Magazine* 15(3):132-255.
1941 Environment and Native Subsistence Economies in the Central Great Plains. *Smithsonian Miscellaneous Collections* 101(3):1-29.
1943 Archaeological Investigations in Platte and Clay Counties, Missouri. *Smithsonian Institution, U.S. National Museum Bulletin 183*. Washington, D.C.
1959 An Introduction to Kansas Archeology. *Smithsonian Institution, Bureau of American Ethnology, Bulletin 174*. Washington, D.C.
1961 *Prehistoric Man on the Great Plains*. University of Oklahoma Press, Norman.
1986 *Central Plains Prehistory: Holocene Environments and Culture Change in the Republican River Basin*. University of Nebraska Press, Lincoln.

Wedel, Waldo R. and Marvin F. Kivett
1956 Additional Data on the Woodruff Ossuary, Kansas. *American Antiquity* 21:414-416.

Wetterstrom, Wilma
1978 Cognitive Systems, Food Patterns, and Paleoethnobotany. In *The Nature and Status of Ethnobotany*, edited by Richard I. Ford, pp.

81-95. Museum of Anthropology, University of Michigan, Anthropological Papers No. 67. Ann Arbor.

WHITAKER, THOMAS W.
1981 Archeological Cucurbits. *Economic Botany* 35:459-466.

WHITAKER, THOMAS W., AND WILLIAM P. BEMIS
1964 Evolution of the Genus Cucurbita. *Evolution* 18:553-559.
1975 Origin and Evolution of the Cultivated *Cucurbita*. *Bulletin of the Torrey Botanical Club* 102:362-368.

WHITAKER, THOMAS W., AND HUGH C. CUTLER
1965 Cucurbits and Cultures in the Americas. *Economic Botany* 19:344-349.
1971 Prehistoric Cucurbits from the Valley of Oaxaca. *Economic Botany* 25(2):123-127.

WILL, GEORGE F., AND GEORGE E. HYDE
1964 *Corn among the Indians of the Upper Missouri*. Reprinted, University of Nebraska Press, Lincoln. Originally published 1917, William H. Miner Co.

WILL, GEORGE F., AND H.J. SPINDEN
1906 *The Mandans*. Papers, Peabody Museum, American Archeology and Ethnography 3, Harvard University.

WILLEY, GORDON R., AND PHILLIP PHILLIPS
1958 *Method and Theory in American Archaeology*. University of Chicago Press, Chicago.

WILLIAMS, BARRY G.
1986 *Early and Middle Ceramic Remains at 14AT2: A Grasshopper Falls Phase House and Pomona Focus Storage Pits in Northeastern Kansas*. Kansas State Historical Society Contract Archeology Publication No. 4 Topeka.

WILSON, GILBERT LIVINGSTON
1977 *Agriculture of the Hidasta Indians, An Indian Interpretation*. Reprinted, J & L Reprint, Lincoln. Originally published 1917, Bulletin of the University of Minnesota, Studies in the Social Sciences, No. 9, Minneapolis.

WING, E.S., AND A.B. BROWN
1979 *Paleonutrition: Method and Theory in Prehistoric Foodways*. Academic Press, New York.

WITTY, THOMAS A., JR.
1962 *Archaeological Investigations of the Hell Creek Valley in the Wilson Reservoir, Russell and Lincoln Counties, Kansas*. Kansas State Historical Society, Anthropological Series, No. 1. Topeka.
1963 *The Woods, Avery, and Streeter Archaeological Sites, Milford Reservoir, Kansas*. Kansas State Historical Society, Anthropological Series No. 2. Topeka.
1966 The West Island Site, 14PH10, A Keith Focus Plains Woodland Site in Kirwin Reservoir, Phillips County, Kansas. *Plains Anthropologist* 11:127-134.
1969 The Caldwell Dig. *Kansas Anthropological Association, Newsletter* No. 15 (2):1-3
1978 Along the Southern Edge: The Central Plains Tradition in Kansas. In *The Central Plains Tradition, Internal Development and External Relationships*, edited by D.J. Blakeslee, pp. 56-66, Report 11, Office of the State Archaeologist, University of Iowa, Iowa City.
1982 *The Slough Creek, Two Dog, and William Young Sites, Council Grove Lake, Kansas*. Kansas State Historical Society, Anthropological Series, No. 10. Topeka.
1983 Personal Communication.

WOOD, C.E.
1977 *Report of Archeological Test Excavations at the proposed Cedar Point Lake, Kansas*. Submitted to the Tulsa District, U.S. Army Corps of Engineers. Wichita State University, Archeology Laboratory, Wichita.

WOOD, W. RAYMOND
1952 The Square Earth Lodges of the Central Great Plains. *Plains Anthropological Conference Newsletter* 5:5-18.
1969a Carbon-14 Dates. In *Two House Sites in the Central Plains: An Experiment in Archaeology*, edited by W. Raymond Wood, pp. 14, Memoir 6, Plains Anthropologist, 14(2).
1969b Ethnographic Reconstructions. In *Two House Sites in the Central Plains: An Experiment in Archaeology*, edited by W. Raymond Wood, pp. 102-108, Memoir 6, Plains Anthropologist, 14(2).

WRIGHT, C.A.
1980 *Archaeological Investigations in the Proposed Blue Springs Lake Area, Jackson County, Missouri: The Early Woodland Period*. Report submitted to Burns and McDonnell Engineers, Kansas City.

WRIGHT, C.M.
1982 An Archaic Site in Osage County, Kansas. *Journal of the Kansas Anthropological Association* 3(1):7-9. Topeka.

YANOVSKY, ELIAS
1936 Food Plants of the North American Indians. United States Department of Agriculture *Miscellaneous Publication* No. 237. Washington, D.C.

YARNELL, RICHARD A.
1964 *Aboriginal Relationships Between Culture and Plant Life in the Upper Great Lakes Region*. Anthropological Papers No. 23, Museum of Anthropology, University of Michigan. Ann Arbor.
1969 Contents of Human Paleofeces. In *The Prehistory of Salts Cave, Kentucky* by Patty Jo Watson et al., pp. 41-54. Illinois State Museum, Reports of Investigations 16, Springfield.

1972 *Iva annua* var. *macrocarpa*: extinct American cultigen? *American Anthropologist* 74:335-341.
1976 Early Plant Husbandry in Eastern North America. In *Cultural Change and Continuity; Essays in Honor of James Bennett Griffin*, edited by Charles E. Cleland, pp. 265-274. Academic Press, New York.
1977 Native Plant Husbandry North of Mexico. In *Origins of Agriculture*, edited by Charles A. Reed, pp. 861-878. Mouton Publishers, The Hague, Paris.
1979 Domestication of Sunflower and Sumpweed in Eastern North America. In *The Nature and Status of Ethnobotany*, edited by Richard I. Ford. Anthropological Papers No. 67, Museum of Anthropology, University of Michigan, Ann Arbor.
1981 Inferred Dating of Ozark Bluff Dweller Occupations Based on Achene Size of Sunflower and Sumpweed. *Journal of Ethnobiology* 1(1):55-60.
1985 Personal Communication.

YELLEN, JOHN E.
1977 *Archaeological Approaches to the Present: Models for Reconstructing the Past*. Academic Press, New York.

ZIEGLER, ROBERT J.
1981a Excavations at 23JA40. In *Prehistoric Cultural Resources Within the Right-of-Way of the Proposed Little Blue River Channel, Jackson County, Missouri*, assembled by Kenneth L. Brown and Robert J. Ziegler. Report on file, U.S. Army Corps of Engineers, Kansas City.
1981b Sites Tested. In *Prehistoric Cultural Resources Within the Right-of-Way of the Proposed Little Blue River Channel, Jackson County, Missouri*, assembled by Kenneth L. Brown and Robert J. Ziegler. Report on file, U.S. Army Corps of Engineers, Kansas City.

Appendix: Summary Data On Measurement Of Cultigens

Table A.1. Summary Data on Measurement of *Cucurbita* and *Lagenaria* seeds.

Cucurbita pepo[a]

Trowbridge (14WY1)

	L[b]		W
	7.8	×	4.6

Hatcher (14DO19)

	L		W
	7.3	×	4.0
	8.0		4.6
	7.8		4.1

N = 3

	L	W
$\bar{X}$ =	7.7	4.2
S =	.36	.32
S^2 =	.08	.07

L/W ratio = .54

Two Deer (14BU55)

	L		W	L		W
	7.9	×	3.8	5.7	×	3.3
	6.9		3.6	8.0		5.4
	10.0		5.2	10.2		6.0
	8.0		4.3	10.8		5.0
	8.2		4.0	7.2		3.5
	12.4		5.7	8.8		4.8
	7.3		4.4	5.9		4.3
	8.2		4.4	7.5		4.0
	6.3		4.2	7.0		4.0
	8.3		4.7	8.0		5.0
	8.2		4.9	7.2		4.2
	6.7		3.6	6.8		3.9
	7.6		3.7			

N = 25

	L	W
$\bar{X}$ =	7.9	4.4
S =	1.5	.71
S^2 =	2.3	.48

L/W ratio = .54

range: 5.7–12.4 3.3–6.0

Friend and Foe (23 CL113)

	L		W
	7.6	×	4.5

Coons (23PL16)

	L		W
	9.4	×	5.0

13ML129

	L	W
Cache 1	7.8	4.4
	10.3	6.0
	7.0	3.8
	8.2	4.4
	7.0	5.0
	6.5	3.3
	6.0	3.6
	7.2	4.7
	7.6	4.3
Cache 5	8.8	3.9
	9.5	5.4
	8.4	4.0
	8.0	3.6
	9.3	4.7
	7.6	4.0
	8.5	4.8
	6.6	3.3
	7.7	4.0
	7.8	3.9
	7.9	3.5
	6.3	3.4

N = 21

	L	W
$\bar{X}$ =	7.8	4.2
S =	1.09	2.04
S^2 =	1.13	.48

L/W ratio = .53

range: 6.0–10.3 3.3–6.0

13ML138

	L	W	L	W
Cache 6	10.2	5.0	6.5	3.0
	8.4	4.4	7.6	4.5
	7.5	3.8	7.5	4.4
	6.2	3.6	7.6	4.2
	9.2	4.5	8.0	5.0
	8.4	4.3	9.5	5.2
	6.2	3.4	6.9	3.6
	9.7	5.3	7.6	4.0
	7.7	4.2	8.0	3.8
	6.0	2.6	7.2	4.4
	7.8	3.7	7.5	3.6
	9.0	5.7	7.1	3.5
	7.8	4.2	6.4	3.3
	8.0	4.5	8.5	3.6
	7.3	4.6	8.5	5.5
	8.2	5.0	9.2	4.4
	9.5	4.6	6.7	3.9
	6.0	4.0	7.6	4.0
	6.8	4.0	10.7	6.8
	6.0	3.2	9.5	4.5
	10.3	5.7	7.3	3.2
	8.0	4.1	6.8	3.2
	9.4	5.0	9.0	4.2
	6.6	2.5	6.3	2.5
	9.4	4.6	7.1	3.1
	5.2	3.3	6.6	3.0
	6.7	3.2	9.9	5.5
	6.7	3.1	7.4	4.8
	7.0	3.0	6.5	4.4
	6.0	4.0	6.5	3.5

Table A.1. Summary Data on Measurement of *Cucurbita* and *Lagenaria* seeds. (continued)

5.7	2.6	5.7	3.1
6.3	3.5	7.0	4.0
7.5	4.5	7.5	4.5
8.2	4.2	6.0	3.5
7.0	4.2	7.0	4.6
7.6	4.3	6.5	4.0
9.0	4.3	8.2	4.4
7.4	4.2	6.2	3.3
7.0	3.2	6.0	3.2
6.5	3.0	8.2	4.1
6.8	4.0	7.6	3.0
8.0	3.7	6.4	3.4
8.7	4.6	7.6	4.7
7.7	3.6	7.8	3.4
7.2	4.2	8.5	5.0
6.2	4.3	6.3	3.0
7.0	3.3	5.4	2.5
6.0	2.8	5.7	3.2
6.2	2.8	8.6	4.4
6.0	2.8	7.5	4.5
6.5	3.1	7.5	4.5
6.5	3.2	9.3	6.0
6.8	3.3	7.6	4.0
8.6	3.4	6.6	3.0
9.4	4.2	7.6	4.0
8.2	4.0	7.0	5.3
7.1	4.0	6.7	3.2

N = 114

	L	*W*
$\bar{X}$ =	7.5	4.0
S =	1.17	.81
S^2 =	1.37	.66

L/W ratio = .53

range: 5.2–10.7 2.5–6.8

Lagenaria siceraria[b]
Coons (23PL16)

L		*W*
9.4	×	4.2
9.1		5.1
9.3		3.8

N = 3

	L	*W*
$\bar{X}$ =	9.3	4.4
S =	.2	.7
S^2 =	.02	.29

[a] All measurements are in mm., no corrections for charring.
[b] Measurements are given only for complete specimens.

Table A.2. Summary Data on Measurements of *Iva annua*.

Trowbridge (14WY1)

L[b]		*W*
4.4	×	3.0
4.9		3.2
4.4		3.2
4.3		3.0
5.2		3.6
4.8		2.9
4.9		2.8
4.5		3.8
5.7		3.3[c]
5.0		4.2
4.6		3.2
5.5		3.0
3.9		3.0
3.9		2.5
4.3		2.0

Kernel
N = 14

	L	*W*	*LW*
$\bar{X}$ =	4.6	3.1	14
S =	.46	.53	
S^2 =	.19	.26	
reconstructed[a]	6.4	4.4	28
range	5.5–7.7	2.8–6.0	

Achene
N = 1

	L	*W*	*LW*
	5.7	3.3	19
reconstructed	6.3	3.7	21

Average $\bar{X}$	6.4	4.1	25
range	5.5–7.7	2.8–6.0	

F test:
- samples compared: L, W measurements from Trowbridge vs. largest of 6 modern collections
- significance level: 5%
- df: $v_1 = 14$; $v_2 = 59$
- critical value: 1.84
- F value: 1.66 (L); 1.52 (W)

T test:
- significance level: 5%
- df: 73
- critical value: 1.65
- T value: 19.8 (L); 9.2 (W)

Young (23PL4)[d]

L[b]		*W*
2.0	×	1.4
2.1		1.7
2.0		1.3
1.6		1.2
3.2		2.1
5.7		3.5
2.4		1.7
1.9		1.2
2.2		1.5
2.2		1.4
2.5		1.6
1.7		1.2
2.0		1.4
1.8		1.3
2.0		1.4

Table A.2. Summary Data on Measurements of *Iva annua*. (continued)

	L	W	LW
	1.8	1.3	
	1.9	1.4	
	2.4	1.4	
	3.2	1.9	
Kernel			
N = 19			
	L	*W*	*LW*
$\bar{X}$ =	2.3	1.6	4
S =	.92	.52	
S^2 =	.80	.26	
reconstructed	3.3	2.3	8
range	2.3–7.9	1.7–5.0	
Fisher-Gabert (23SA128)			
	L	*W*	
	5.2 ×	3.3	
	3.8	2.3	
Kernel			
N = 2			
	L	*W*	*LW*
$\bar{X}$ =	4.5	2.8	13
S =	.98	.70	
S^2 =	.49	.25	
reconstructed	6.3	4.0	25
Two Deer (14BU55)			
	L	*W*	
	4.9 ×	3.5[c]	
	4.7	3.1[c]	
	4.6	3.4[c]	
	4.3	3.0[c]	
	4.7	2.4	
	5.0	3.0	
	4.4	2.6	
	3.0	2.3	
	4.2	2.8	
	4.3	2.6	
	3.8	3.0	
	4.3	3.5	
	5.2	3.2	
	4.6	3.5	
	5.3	3.0	
	4.4	3.3	
	4.2	2.3	
	4.2	3.3	
	4.1	2.8	
	5.1	3.2	
	4.7	2.4	
Kernel			
N = 17			
	L	*W*	*LW*
$\bar{X}$ =	4.4	2.9	13
S =	.56	.40	
S^2 =	.29	.16	
reconstructed	6.2	4.1	25
range	4.3–7.4	3.3–5.0	
Achene			
N = 4			
	L	*W*	*LW*
$\bar{X}$ =	4.6	3.3	15
S =	.25	.23	
S^2 =	.04	.04	
reconstructed	5.1	3.7	19
range	4.7–5.4	3.3–3.9	
Average $\bar{X}$	5.7	3.9	22
range	4.3–7.4	3.3–5.0	

F test:
samples compared: length measurements from Two Deer and largest of 6 modern samples
significance level: 5%
df: $v_1 = 20$; $v_2 = 58$
critical value: 1.75
F value: 2.05

	L	W	LW
23CL274			
	L	*W*	
	3.9 ×	2.3	
	4.7	2.7	
	4.2	2.8	
	4.1	2.7	
	3.9	2.1	
	4.2	2.8	
	4.0	2.3	
	4.3	2.2	
Kernel			
N = 8			
	L	*W*	*LW*
$\bar{X}$ =	4.2	2.5	11
S =	.26	.29	
S^2 =	.05	.07	
reconstructed	5.9	3.6	21
range:	5.5–6.6	2.9–4.0	

T test:
samples compared: L measurements from 23CL274 vs. largest of 6 modern samples
significance level: 5%
df: 66
Critical value: 1.66
T value: 7.33 (L)

	L	W	LW
Friend and Foe (23CL113)[e]			
	L	*W*	
	4.5 ×	?	
	*4.1	3.8	
	*5.0	2.6	
	*5.6	3.7	
	4.9	3.3	
	4.2	2.9	
Kernel			
N = 9			

*Note: These three measurements represent the ranges in size for six seeds (Calabrese 1969:158). Individual measurements for all recovered seeds are not provided. Assuming that the measurements listed are from seeds, not achenes, size corrections were calculated.

	L	W	LW
$\bar{X}$ =	4.7	3.3	16
S =	.56	.51	
S^2 =	.26	.21	
corrected	6.6	4.7	31
range	5.7–7.8	3.7–5.5	

Table A.2. Summary Data on Measurements of *Iva annua*. (continued)

23CL115[f]

L		W
5.1	×	3.6[c]
3.2		2.4
3.7		2.5
3.8		2.3
3.8		2.6
3.8		2.8
4.0		2.6
4.1		2.9
4.1		2.9
4.2		2.8
4.2		3.8
4.3		2.6
4.3		2.7
4.4		3.0
4.5		2.9
4.6		2.3
4.9		3.4

Kernel
N = 16

	L	W	LW
$\bar{X}$ =	4.1	2.8	11
S =	.40	.39	
S^2 =	.15	.14	
reconstructed	5.7	4.0	23
range	4.5–6.8	3.4–4.9	

Achene
N = 1

	L	W	LW
	5.1	3.6	
reconstructed	5.6	4.0	
Average X	5.7	4.0	23
range	4.5–6.8	3.4–4.9	

T test:
samples compared: L,W measurements from 23CL115 vs. largest of 6 modern samples
significance level: 5%
df: 75
critical value: 1.66
F value: 17.3 (L); 7.27 (W)

Young (23PL4)[f,g]

L		W
4.4	×	2.9
4.2		2.3
3.4		2.2
4.8		2.8

Kernel
N = 4

	L	W	LW
$\bar{X}$ =	4.2	2.6	11
S =	.59	.35	
S^2 =	.26	.09	
reconstructed	5.9	3.7	22
range	4.8–6.7	3.1–4.2	

T test:
samples compared: L,W measurements from Young vs. largest of 6 modern samples
significance level: 5%
df: 62
critical value: 1.67
t value: 7.14 (L); .86 (W)

14ML417

L		W
4.3	×	2.6
4.5		2.5
4.5		2.3
4.6		3.2
3.7		2.9

Kernel
N = 5

	L	W	LW
$\bar{X}$ =	4.3	2.7	12
S =	.36	.35	
S^2 =	.10	.1	
reconstructed	6.0	3.9	23
range:	5.2–6.4	3.3–4.6	

T test:
samples compared: L,W measurements from 14ML417 vs. largest of 6 modern samples
significance level: 5%
df: 63
critical value: 1.66
T value: 8.9 (L); 3.63 (W)

13ML129
Cache 1

	L	W	L	W
	4.6	2.4	6.5	3.3[c]
	4.6	3.2	4.8	3.0[c]
	4.2	3.0	4.9	3.0[c]
	5.0	3.0	4.0	2.6[c]
	4.5	3.0	4.3	3.0[c]
	3.7	2.5	4.5	2.6[c]
	4.6	3.0	4.5	2.5[c]
	4.2	2.9	4.0	3.0[c]
	4.5	3.0	4.8	2.3[c]
	4.9	3.0	4.7	3.0[c]
	3.5	2.4	5.5	4.0[c]
	4.5	3.0	5.3	3.5[c]
	4.6	3.3	4.0	2.7[c]
	4.8	2.8	5.0	3.5[c]
	4.6	3.0	4.0	2.0[c]
	4.2	2.5	4.5	2.7[c]
	4.0	3.4	4.6	3.1[c]
	4.8	3.2	6.3	3.7[c]
	5.0	4.0	5.4	4.0[c]
	4.5	2.5	4.0	3.2[c]
	4.6	2.6	4.0	3.1[c]
	4.6	3.1	4.5	2.6[c]
	5.0	3.9[c]	4.8	3.0[c]
	3.7	3.8[c]	4.5	3.0[c]
	4.0	3.2[c]	4.8	3.2[c]
	5.2	3.0[c]	3.7	2.8[c]
	5.6	3.1[c]	3.9	3.0[c]
	3.8	3.4[c]	3.5	2.5[c]
	4.3	2.9[c]	4.7	2.5[c]
	4.0	3.1[c]	4.5	2.5[c]
	3.5	3.4[c]	4.5	3.0[c]
	4.6	3.5[c]		
323–710	6.0	4.0		
Cache 5	6.0	3.3	6.3	4.6[c]
	4.6	3.0	6.8	4.3[c]

Table A.2. Summary Data on Measurements of *Iva annua*. (continued)

5.0	3.6	5.6	3.9[c]
4.7	3.5	4.4	3.2[c]
4.5	2.5	6.0	4.0[c]
4.7	3.3	4.8	3.5[c]
4.5	4.0	5.9	4.5[c]
4.4	4.0	4.7	3.9[c]
5.1	4.0	5.6	3.4[c]
4.5	3.9	5.5	3.9[c]
5.0	3.6	5.8	3.8[c]
5.0	3.8	6.7	4.2[c]
5.0	2.9	6.2	3.9[c]
5.5	3.0	6.0	4.3[c]
6.4	4.4	5.7	3.6[c]
4.5	2.8	5.5	3.5[c]
3.8	3.4	6.5	4.2[c]
3.8	2.7	6.5	3.2[c]
4.7	3.2	7.0	3.3[c]
5.5	3.1	5.0	3.6[c]
5.0	3.0	5.9	3.6[c]
4.0	3.4	6.0	4.3[c]
3.2	3.3	6.0	4.5[c]
3.6	5.3	6.0	4.0[c]
4.7	4.0	6.0	4.1[c]
4.8	3.1	6.2	3.9[c]
5.2	3.1	4.6	3.1[c]
4.3	3.0	5.0	3.0[c]
4.3	3.7	6.0	3.2[c]
4.0	2.1	6.0	3.5[c]
4.8	2.4	4.7	3.4[c]
5.2	3.5	5.7	3.8[c]
4.5	2.5	7.4	4.3[c]
4.3	2.6	5.7	3.5[c]
4.3	3.2	6.0	3.5[c]
5.1	2.5	6.3	4.0[c]
4.4	2.6	5.8	5.0[c]
5.5	3.7	5.4	3.7[c]
4.5	3.2	6.0	3.4[c]
5.3	3.4	6.0	4.4[c]
4.4	3.9	6.4	3.9[c]
5.5	3.5	6.2	3.7[c]
4.2	3.4	5.5	4.1[c]
4.2	3.5	4.3	2.0
5.0	2.5	4.5	3.5
4.5	3.2	5.0	3.2
4.5	3.3	4.5	2.6
3.0	2.6	4.7	3.5
3.3	2.6	5.2	3.9
3.9	2.5	4.6	2.5
3.7	3.3	3.7	2.5
4.1	2.6	5.0	4.3
5.0	4.0	4.5	3.3
4.2	3.0	5.8	3.6
5.0	3.4	4.5	2.8
4.2	4.0	4.7	3.0
4.2	3.2	4.6	3.0
4.2	3.6	5.0	3.8
5.2	3.7	4.8	3.2
4.9	3.0	4.3	3.5
5.5	3.1	4.2	2.8
4.3	3.3	4.4	2.5
4.0	3.2	5.0	3.7
4.7	3.2	4.9	3.3

4.3	3.4	4.8	2.4
4.3	3.5	4.0	3.1
4.0	3.9	4.1	3.9
4.6	3.5	5.0	3.6
4.5	3.4	5.2	3.0
4.7	3.3	4.5	2.3
5.3	2.8	3.5	2.8
5.0	2.7	4.0	2.7
4.8	2.9	4.5	2.9
5.0	3.4	5.2	4.0
5.2	3.0	3.3	2.4
4.3	2.5	4.2	3.0
4.5	3.5	4.7	4.2
5.0	3.0	3.9	2.8
4.2	3.4	3.6	3.7
5.0	3.2	4.5	2.8
5.0	3.5	4.5	3.7
5.7	3.0	5.2	3.5
5.7	3.5	3.7	2.6
4.7	3.6	5.5	3.7
4.8	2.5	5.7	3.5
4.5	2.7	4.5	3.0
4.5	2.8	4.8	2.3
5.0	4.0	4.5	3.0
4.8	3.8	4.4	3.5
4.5	2.0	4.5	3.5
5.2	3.0	6.2	4.0
5.0	3.9	4.5	3.7
5.0	3.5	4.5	3.1
4.8	2.8	4.2	3.9
5.2	3.5	5.2	3.8
6.0	3.5	5.1	3.2
5.0	3.4	5.0	3.2
5.0	3.5	4.4	3.2
3.7	2.3	3.5	2.4
5.2	2.7	4.3	3.3
5.5	3.4	4.8	5.0
4.4	3.4	4.7	3.0
5.2	3.3	3.8	2.9
4.5	3.0	4.9	3.8
4.7	2.0	5.8	3.5
4.5	3.0	3.5	3.2
4.7	2.5	4.0	2.8
5.5	3.2	5.0	3.1
4.8	3.4	3.5	2.7
6.0	4.0	4.5	3.3
3.0	2.0	5.4	3.2
6.3	3.9	4.2	3.9
4.8	3.3	5.4	3.0
5.4	3.6	4.0	3.6
3.6	2.1	5.3	3.1
4.4	3.3	4.5	2.7
4.5	3.5	4.9	3.2
4.6	3.1	4.4	3.5
5.1	3.6	4.8	3.6
4.8	3.4	3.9	2.5
5.0	3.0	5.0	3.0
4.6	3.3	4.0	2.3
4.0	3.5	4.6	2.5
4.0	3.0	3.9	2.8
5.6	3.7	3.6	2.6
5.4	3.8	4.5	3.2
4.8	3.5	5.2	3.0
5.0	3.0	5.2	4.3

Table A.2. Summary Data on Measurements of *Iva annua*. (continued)

	5.4	3.5		4.2	3.0
	4.6	4.0		4.0	3.0
	4.5	2.3		4.8	3.5
	6.9	3.2		4.3	3.3
	4.5	2.5		4.0	3.2
	4.5	3.5		4.6	3.6
	4.2	3.0		4.4	2.6
	4.2	3.5		4.5	3.2
	4.5	3.7		4.6	3.0
	5.5	4.0		5.0	2.6
	4.0	2.9		4.5	2.7
	5.2	4.0		4.3	3.2
	4.5	3.5		5.3	3.1
	4.6	3.0		4.9	3.6
	4.0	2.5		5.0	3.9
	4.6	2.6		4.8	2.8
	5.0	4.1		3.5	2.5
	4.7	3.5		3.3	2.4
	4.2	3.2		5.0	3.4
	6.0	3.4		3.9	3.4
	4.0	2.9		4.5	3.0
	4.8	3.1		5.0	3.0
	4.4	2.6		4.0	2.8
	5.0	3.2		4.5	3.3
	4.5	3.2		4.5	3.2
	4.5	3.4		3.8	3.3
	5.0	2.5		4.0	2.3
	4.3	2.8		3.6	2.6
	4.7	3.9		5.5	3.2
	4.4	3.9		4.2	3.3
	5.6	4.0		3.1	2.0
	4.1	3.4		4.2	2.2
	4.3	2.3		5.0	3.0
	3.7	2.8		3.8	2.4
	4.0	2.7		4.3	3.0
	2.7	2.5		3.6	2.5
Kernel					
N = 308					
	L	*W*	*LW*		
$\bar{X}$ =	4.6	3.2	15		
S =	.66	.52			
S^2 =	.43	.27			
reconstructed	6.4	4.9	31		
range	3.8–9.5	2.8–6.3			
Achene					
N = 84					
	L	*W*	*LW*		
$\bar{X}$ =	5.2	3.5	18		
S =	.92	.59			
S^2 =	.85	.35			
reconstructed	5.8	3.9	23		
range	3.8–8.6	1.9–5.1			
Average $\bar{X}$	6.1	4.4	27		
range	3.8–9.5	1.9–6.3			
13ML138					
Cache 6					
	L	*W*		*L*	*W*
	5.0	3.3		5.0	3.3[c]
	4.2	2.6		5.2	3.0[c]
	5.7	3.0		6.6	3.2[c]
	4.0	3.0		5.5	3.5[c]
	4.5	2.8		4.2	2.8[c]
	4.0	1.7[c]		3.7	2.7[c]
Kernel					
N = 5					
	L	*W*	*LW*		
$\bar{X}$ =	4.7	2.9	14		
S =	.68	.26			
S^2 =	.37	.05			
reconstructed	6.5	4.1	27		
range	5.6–7.9	4.0–4.7			
Achene					
N = 7					
	L	*W*	*LW*		
$\bar{X}$ =	4.9	2.9	14		
S =	1.0	.59			
S^2 =	.87	.30			
reconstructed	5.4	3.2	17		
range	4.5–7.3	1.9–3.9			
Average $\bar{X}$	6.0	3.7	22		
range	4.5–7.9	1.9–4.7			
Witt (14GE600)					
	L	*W*		*L*	*W*
	4.8	3.0		4.2	3.4
	5.6	3.8[c]		5.0	3.5
	4.2	3.0		4.3	2.5
	5.6	3.4		3.5	2.6
Kernel					
N = 7					
	L	*W*	*LW*		
$\bar{X}$ =	4.5	3.1	14		
S =	.67	.39			
S^2 =	.39	.13			
reconstructed	6.3	4.4	28		
range	4.9–7.8	3.6–5.0			
Achene					
N = 1					
	L	*W*	*LW*		
reconstructed	6.2	4.2	26		
Average $\bar{X}$	6.3	4.3	27		
range	4.9–7.8	3.6–5.0			

[a] adjusted according to Asch and Asch (1985) for loss of pericarp and shrinkage.

[b] all measurements are in mm.

[c] complete achene; size adjusted for shrinkage only.

[d] Kansas City Hopewell occupation.

[e] corrections for charring were calculated, assuming that the measurements provided by Calabrese (1969:158) represent uncorrected seeds. Statistics were calculated on this sample even though some measurements represent ranges, not individual seed sizes. However, no comparisons were made to modern samples.

[f] measurements provided by Leonard Blake, Missouri Botanical Garden, in a letter to P. J. O'Brien, 1974.

[g] Steed Kisker occupation.

Table A.3. Summary Data on Measurement of *Helianthus annuus*.

Coffey (14PO4) Horizon III-3

L^{b}		W
4.3	×	2.2
3.3		2.1
3.2		1.6
4.4		2.2
4.0		1.7
4.2		1.8

Achene
N = 6

	L	*W*	*LW*
$\bar{X}$ =	3.9	1.9	8
S =	.52	.26	
S^2 =	.23	.06	
reconstructed[c,a]	4.3	2.5	11
range	3.5–4.8	2.0–2.8	

Hatcher (14DO19)

L		*W*
8.9	×	3.6
8.0		3.0
8.6		3.7
7.2		2.2
6.9		3.2
7.4		3.1
8.3		2.8
7.3		3.5

Kernel
N = 8

	L	*W*	*LW*
$\bar{X}$ =	7.8	3.1	24
S =	.72	.49	
S^2 =	.46	.20	
reconstructed	10.1	4.5	46
range	9.0–11.6	3.2–5.4	

F test:
samples compared: L, W measurements from Hatcher vs. largest of 2 modern samples
significance level: 5%
df: $v_1 = 7$; $v_2 = 155$
critical value: 2.64
F value: 13.0 (L); 6.4 (W)

Two Deer (14BU55)

L	*W*	*L*	*W*
7.0	3.4	7.3	2.8[d]
7.6	4.2	8.0	3.8[d]
7.0	2.7	6.7	4.0[d]
7.6	3.1	8.0	3.7[d]
9.0	3.7	7.8	3.2[d]
8.8	3.0	10.3	4.8[d]
7.0	3.2	8.9	3.0[d]
7.3	3.2	9.0	3.0[d]
6.5	3.0	7.8	3.0[d]
7.0	2.6	7.5	3.3[d]
6.5	2.9	8.0	3.5[d]
7.0	2.8	6.1	2.5[d]
7.6	2.8	6.0	2.6[d]
7.0	2.6	7.8	3.0[d]
6.4	2.0	5.9	2.8[d]
6.9	2.5	7.5	2.9[d]
7.5	3.3	7.3	2.7[d]
9.3	3.6	8.0	3.5[d]
7.8	2.8	8.0	3.7[d]
7.4	2.6	7.6	3.3[d]
7.3	2.8	7.5	3.8[d]
7.1	2.8	6.5	3.0[d]
6.9	2.8	13.8	5.7[d]
8.0	3.0	7.5	3.5[d]
7.7	3.7	8.0	2.8[d]
6.2	2.7	5.9	2.9[d]
7.3	3.0	7.0	2.2[d]
7.8	3.0	7.2	3.4[d]
7.3	2.5	8.2	3.5[d]
6.9	2.7	9.5	5.0[d]
6.1	2.6	6.2	2.8[d]
8.5	3.2	6.5	2.7[d]
7.4	2.8	7.8	3.9[d]
7.3	2.2	8.9	3.5[d]
7.0	2.8	7.9	2.6[d]
6.6	2.5	6.0	2.6[d]
7.5	2.6	7.5	3.4[d]
7.8	2.5	7.0	3.0[d]
9.3	3.1	9.0	3.5[d]
8.0	3.7	8.0	3.5[d]
7.8	2.2	8.0	3.2[d]
7.6	2.5	8.2	3.0[d]
6.0	3.1	8.0	2.7[d]
8.2	3.3	7.3	4.2[d]
6.5	2.9	7.5	2.4[d]
8.5	3.0	7.6	3.5[d]
8.1	3.0	7.2	2.6[d]
8.2	3.0	9.0	3.5[d]
8.1	3.0	7.5	3.5[d]
7.0	2.5	8.5	4.0[d]
7.5	2.6	8.5	2.5[d]
5.7	2.8[d]	8.2	3.3[d]
6.5	2.7[d]	8.9	3.3[d]
6.0	2.0[d]	7.5	3.0[d]
6.8	2.7[d]	8.2	4.2[d]
7.7	2.7[d]	6.3	2.4[d]
8.1	4.2[d]	7.4	2.8[d]
7.0	3.2[d]	7.8	2.4[d]
8.0	3.8[d]	6.9	2.7[d]
8.7	2.8[d]	7.2	2.6[d]

Kernel
N = 51

	L	*W*	*LW*
$\bar{X}$ =	7.4	3.2	24
S =	.76	1.99	
S^2 =	.57	3.91	
reconstructed	9.6	4.6	45
range	7.8–12.1	2.9–6.1	

Achene
N = 69

	L	*W*	*LW*
$\bar{X}$ =	7.7	3.2	25
S =	1.19	.66	
S^2 =	1.40	.43	
reconstructed	8.5	4.1	35
range	6.5–15.3	2.5–7.2	
Average $\bar{X}$	9.1	4.4	40

Table A.3. Summary Data on Measurement of *Helianthus annuus*. (continued)

	L	W
range	6.5–15.3	2.5–7.2

F test:
samples compared: Length Width and L×W measurements from Two Deer vs. modern El Dorado sample
significance level: 5%
df: $v_1 = 119$; $v_2 = 155$
critical value: 1.46
F value: 17.5 (L); 8.2 (W); 50.9 (LW)

Young (23PL4)[d]

L		*W*
9.0	×	3.7
9.2		3.4
8.3		3.2
7.5		3.4
7.7		3.3
8.5		3.0
9.5		3.2
7.9		3.2
7.8		3.3
8.2		2.5
7.8		3.2
7.2		2.5
8.2		3.1
6.2		2.5
6.5		2.5
6.1		2.2
7.0		2.8
7.0		2.9
9.0		3.5
6.6		2.8
7.3		2.8
5.9		2.5
5.7		2.6
4.9		2.0
5.7		2.6
5.5		2.8
5.2		2.4
5.2		2.4
5.2		2.0
5.0		2.0
5.2		2.0
5.0		2.0
3.9		2.2

Achene
N = 33

	L	*W*	*LW*
$\bar{X}$ =	6.8	2.7	18
S =	1.5	.49	
S² =	2.2	.24	
reconstructed	8.8	3.9	34
range	5.1–12.4	2.9–5.4	

F test:
samples compared: L, W, LW from Young vs. largest of modern samples
significance level: 5%
df: $v_1 = 32$; $v_2 = 67$
critical value: 1.88
F value: 23.6 (L); 7.4 (W); 42 (LW)

14BU71

	L		*W*	*LW*
	7.2	×	2.7	
reconstructed	9.4		3.9	37

14ML417

L		*W*
8.0	×	3.1
4.3		1.9
6.9		2.6

Kernel
N = 3

	L	*W*	*LW*
$\bar{X}$ =	6.4	2.5	16
S =	1.9	.60	
S² =	2.4	.24	
reconstructed	8.3	3.6	30
range	5.6–10.4	2.8–4.5	

F test:
samples compared: L, W measurements from 14ML417 vs. largest of modern samples
signifcance level: 5%
df: $v_1 = 2$; $v_2 = 155$
critical value: 3.0
F value: 72.5 (L); 7.6 (W)

Anderson (14CF508)

L		*W*
5.0	×	2.2
5.1		2.3
5.3		1.8
5.6		1.8
5.9		2.0

Kernel
N = 5

	L	*W*	*LW*
$\bar{X}$ =	5.4	2.0	11
S =	.37	.23	
S² =	.10	.04	
reconstructed	7.0	2.9	20
range	6.5–7.7	2.6–3.3	

T test:
samples compared: L, W measurements from Anderson vs. largest of modern collection
significance level: 5%
df: 159
critical value: 1.64
T value: 13.3 (L); 5.0 (W)

14MN509

	L		*W*	*LW*
	10.1	×	3.7	
reconstructed	13.1		5.4	70

13ML129

	L	W	L	W
Cache 1	9.5	3.0	6.9	3.0[d]
	7.8	1.8	5.6	2.8[d]
	9.0	4.0	5.0	2.7[d]
	7.6	3.8	8.6	2.8[d]
	9.0	3.3	7.6	2.5[d]
	9.3	3.4	6.0	2.3[d]
	7.8	2.6	6.4	2.4[d]

Table A.3. Summary Data on Measurement of *Helianthus annuus*. (continued)

	6.0	2.5	7.0	3.7[d]
	9.5	3.5	8.3	2.4[d]
	9.3	3.0	7.0	2.0[d]
	8.8	3.6	5.0	2.3[d]
	8.0	3.5	8.7	3.2[d]
	7.6	3.8	7.1	3.2[d]
	8.0	3.7	6.3	2.2[d]
	6.0	2.5	5.5	2.4[d]
	7.5	2.5	8.5	3.0[d]
	6.7	2.8	5.2	2.3[d]
	7.5	2.6	11.5	4.0[d]
	6.8	2.8	6.0	2.0[d]
	8.6	3.2	5.5	2.0[d]
	6.3	2.3	8.5	2.5[d]
	8.0	3.5	6.3	2.8[d]
	7.8	3.6	5.5	2.2[d]
	6.5	2.6	5.0	2.1[d]
	9.0	3.0	5.0	2.4[d]
	8.0	3.0	8.8	3.2[d]
	6.0	2.2	5.7	2.2[d]
	7.8	2.6[d]	7.7	2.6[d]
	7.1	2.6[d]	6.6	2.5[d]
	6.7	1.8[d]	5.2	2.3[d]
	5.3	2.2[d]	7.0	2.4[d]
	6.0	2.5[d]	7.5	3.0[d]
	7.0	2.2[d]	5.3	1.7[d]
	8.0	2.3[d]	8.4	3.0[d]
	6.8	3.1[d]	8.5	3.3[d]
	5.5	2.3[d]	6.6	2.4[d]
	7.1	2.7[d]	4.7	2.2[d]
	6.9	2.6[d]	6.0	2.2[d]
	6.3	2.5[d]	7.8	2.6[d]
	9.0	2.7[d]	8.5	2.7[d]
	5.4	2.6[d]	7.7	3.5[d]
	9.0	3.0[d]	8.3	3.4[d]
	6.0	2.4[d]	7.5	3.5[d]
	8.7	3.3[d]	12.2	3.6[d]
	10.0	3.5[d]	7.8	3.5[d]
	6.5	1.9[d]	8.8	3.7[d]
	6.0	2.6[d]	5.7	2.2[d]
	7.3	2.2[d]	6.8	3.5[d]
	8.0	2.0[d]	7.8	3.7[d]
	5.9	2.3[d]	5.5	2.0[d]
	12.0	3.0[d]	6.0	2.5[d]
	7.2	4.5[d]	6.0	2.8[d]
	5.4	2.5[d]	7.5	3.5[d]
Cache 5	11.0	4.0	5.8	2.0[d]
	6.5	2.3	6.5	2.3[d]
	7.5	3.5	6.0	2.3[d]
	7.5	3.4	10.8	3.2[d]
	9.0	3.0	7.5	3.0[d]
	7.7	3.0	11.5	3.4[d]
	9.2	2.5	9.0	3.5[d]
	7.6	3.1	9.5	3.0[d]
	10.0	3.5	6.5	2.8[d]
	5.8	2.0	8.2	2.5[d]
	8.0	2.7	7.0	2.8[d]
	8.4	2.6	6.3	2.8[d]
	6.0	2.1	7.0	2.3[d]
	10.5	4.2	7.6	2.7[d]
	9.3	3.0	6.4	2.5[d]
	7.7	3.5	7.5	3.5[d]
	10.6	3.8	6.0	3.0[d]
	6.4	2.8	8.5	2.4[d]
	10.0	3.0	8.7	2.6[d]
	7.3	2.7	7.8	3.2[d]
	9.4	4.2	5.7	2.0[d]
	7.6	2.1	5.8	2.5[d]
	7.0	2.2	8.0	3.5[d]
	8.0	3.6	6.1	2.4[d]
	8.4	2.4	6.3	2.8[d]
	9.5	2.9	8.5	3.8[d]
	5.5	1.8	5.7	2.5[d]
	7.9	2.0	10.2	3.5[d]
	8.8	2.4	7.7	2.0[d]
	7.0	2.2	7.6	3.0[d]
	6.0	2.0	9.6	2.6[d]
	8.5	3.3	7.5	3.2[d]
	7.1	2.4	6.4	2.3[d]
	7.2	2.9	8.0	2.7[d]
	10.0	4.2	10.0	3.4[d]
	6.6	1.9	6.8	2.3[d]
	7.5	2.0	5.8	2.0[d]
	9.0	2.8	6.3	2.4[d]
	6.9	2.0	8.0	2.6[d]
	8.0	2.5	7.0	2.4[d]
	5.5	1.7	11.1	3.6[d]
	9.3	2.6	9.0	2.5[d]
	9.0	3.5	6.1	1.7[d]
	6.3	2.2	7.0	2.6[d]
	6.4	2.1	7.4	3.9[d]
	7.8	2.9	8.7	2.6[d]
	6.4	2.5	9.3	3.6[d]
	7.2	2.4	9.0	2.8[d]
	8.0	3.0	8.5	3.5[d]
	6.5	2.0	6.0	2.6[d]
	7.0	2.0	9.9	3.8[d]
	6.6	2.4	9.0	3.0[d]
	8.1	3.0	8.5	3.6[d]
	7.3	2.4	6.0	2.8[d]
	5.7	2.3	5.3	2.3[d]
	8.4	1.8	5.4	2.1[d]
	8.6	2.9	6.0	2.8[d]
	6.1	2.3	7.0	2.5[d]
	7.8	2.8	7.0	1.8[d]
	7.3	3.1	8.0	3.0[d]
	8.3	2.5	7.6	2.5[d]
	6.8	2.0	6.2	2.8[d]
	9.2	3.3	9.7	2.7[d]
	11.2	3.8	7.8	2.4[d]
	10.7	4.2	8.4	3.4[d]
	6.8	2.2	6.8	3.0[d]
	9.8	3.2	8.2	2.4[d]
	6.7	2.0	10.2	3.8[d]
	6.5	2.5	6.8	2.7[d]
	9.5	3.4	7.7	2.6[d]
	7.2	2.9	8.9	2.9[d]
	6.0	2.2	8.3	2.6[d]
	9.3	2.7	9.8	2.7[d]
	4.7	1.8	6.1	2.9[d]
	5.5	2.1	8.2	2.9[d]
	9.0	3.0	7.0	2.8[d]
	8.1	3.5	7.5	2.5[d]
	7.7	2.5	4.8	2.2[d]
	7.8	2.7	7.1	2.4[d]
	7.5	2.4	6.0	2.0[d]

Table A.3. Summary Data on Measurement of *Helianthus annuus*. (continued)

	5.1	2.0	11.2	2.8[d]
	5.2	2.2	6.0	2.2[d]
	4.7	1.9	6.3	3.0[d]
	7.2	2.7	9.7	2.5[d]
	5.5	1.8[d]	7.5	1.8[d]
	7.1	2.5[d]	7.1	2.5[d]
	5.9	3.0[d]	9.7	3.8[d]
	7.8	2.5[d]	6.2	2.5[d]
	6.7	2.5[d]	6.4	3.4[d]
	7.2	2.7[d]	7.4	3.0[d]
	7.3	2.9[d]	7.6	2.0[d]
	8.5	2.5[d]	7.0	2.3[d]
	4.6	1.7[d]	7.0	2.9[d]
	7.3	2.4[d]	6.2	2.5[d]
	5.5	2.0[d]	5.8	1.8[d]
	9.2	3.3[d]	10.0	3.3[d]
	6.7	3.0[d]	6.8	3.3[d]
	9.2	3.5[d]	6.5	2.0[d]
Kernel				
N = 113				
	L	*W*	*LW*	
$\bar{X}$ =	7.8	2.8	22	
S =	1.39	.65		
S^2 =	1.92	.43		
reconstructed	10.1	4.1	41	
range	6.6–14.6	2.5–6.1		
Achene				
N = 193				
	L	*W*	*LW*	
$\bar{X}$ =	7.4	2.7	20	
S =	1.57	.54		
S^2 =	2.46	.29		
reconstructed	8.2	3.4	28	
range	5.1–13.5	2.2–5.7		
Average $\bar{X}$	9.2	3.8	35	
range	5.1–14.6	2.2–6.1		
13ML138				
Cache 6	*L*	*W*	*L*	*W*
	12.0	4.5	10.5	4.3
	9.2	3.0	9.8	3.3
	7.5	2.0	6.8	2.5
	8.2	2.0	7.8	3.0
	7.4	2.0	8.5	3.8
	7.8	2.3	5.7	2.4
	5.5	1.7	7.5	2.5
	8.4	3.0	7.0	2.7[d]
	9.2	2.5[d]	8.2	3.4[d]
	5.5	2.3[d]	8.8	3.5[d]
	8.3	3.0[d]	8.5	3.5[d]
Kernel				
N = 8				
	L	*W*	*LW*	
$\bar{X}$ =	8.3	2.6	22	
S =	1.85	.92		
S^2 =	3.0	.74		
reconstructed	10.7	3.8	41	
range	7.2–15.6	2.5–6.5		
Achene				
N = 14				
	L	*W*	*LW*	
$\bar{X}$ =	8.0	3.1	25	
S =	1.42	.60		
S^2 =	1.9	.33		
reconstructed	8.9	3.9	35	
range	6.1–11.6	3.1–5.5		
Average $\bar{X}$	9.8	3.9	38	
range	6.1–15.6	2.5–6.5		
Witt (14GE600)				
	L	*W*		
	6.6	3.3		
	7.0	3.6		
	7.7	3.5		
Kernel				
N = 3				
	L	*W*	*LW*	
$\bar{X}$ =	7.1	3.5	25	
S =	.55	.15		
S^2 =	.21	.01		
reconstructed	9.2	5.1	47	

[a] corrections according to Yarnell 1985.

[b] all measurements are in mm for complete specimens only.

[c] sample does not represent a cultivated variety.

[d] complete achene; size adjusted for shrinkage only.

[e] measurements provided by Leonard Blake, Missouri Botanical Garden in letters to P. J. O'Brien, 1974 and 1980. Sample does not include specimens reported in Yarnell (1979).

Table A.4. Summary Data on Measurements of *Zea mays* kernels and cupules.

Trowbridge (14WY1)[a]

L[b]	W	T	L	W	T
9.0	7.0	4.0	7.8	7.5	3.5
7.5	6.5	3.0	7.5	7.2	2.8
8.0	7.5	4.3	8.0	6.8	3.0
8.0	7.3	2.7	7.2	6.5	4.3
8.3	7.5	3.8	6.8	6.5	4.0
9.0	6.5	4.0	8.0	7.0	–
8.0	7.0	3.7	9.0	9.0	5.5
8.2	6.5	3.4	8.2	8.5	3.6
9.0	7.5	3.6	8.0	6.8	3.8
8.5	7.3	3.4	7.8	6.7	3.6
8.6	6.7	3.6	8.5	6.8	3.2
8.3	6.7	3.0	8.5	7.0	3.2
9.2	7.0	3.7	7.0	6.8	3.4
8.0	7.0	3.3	7.7	6.8	3.0
8.5	6.5	3.9	8.0	8.0	3.2
6.8	7.5	–	8.5	6.5	3.5
8.2	7.4	4.4	7.6	7.0	3.7
8.0	6.8	3.2	7.0	7.8	3.6
8.0	7.0	3.7	7.6	6.8	3.2
8.0	8.3	3.4	7.6	7.0	4.4
8.5	7.0	3.3	7.8	7.2	3.2

N = 42

	L	W	T	W/L	T/W
X̄ =	8.2	7.1	3.6	.86	.51
S =	.6	.6	.5		
S^2 =	.32	.31	.26		
range:	7.0–9.2	6.5–9.0	2.7–5.5		

Two Deer (14BU55)

L	W	T	L	W	T
6.3	7.2	5.4	5.0	6.8	3.5
6.5	8.5	4.2	6.0	6.4	4.4
5.5	7.0	4.3	6.3	6.5	3.7
5.4	6.6	3.9	6.0	7.0	5.2
5.0	6.5	5.0	6.7	7.5	4.2
4.8	6.5	4.4	5.4	6.5	3.9
5.5	6.8	3.4	6.3	7.8	5.4

N = 14

	L	W	T	W/L	T/W
X̄ =	5.8	7.0	4.4	.82	.62
S =	.6	.6	.7		
S^2 =	.35	.34	.42		
range:	4.8–6.7	6.5–8.5	3.4–5.4		

14BU57

L	W	T
6.5	7.2	4.0

Anderson (14CF508)

L	W	T
8.2	9.0	4.6
4.6	3.8	3.0
4.7	3.5	3.2
5.6	3.7	3.4
5.4	3.9	3.4

N = 5

	L	W	T	W/L	T/W
=	5.7	4.8	3.5	.84	.73
S =	1.5	2.4	.6		
S^2 =	1.7	4.5	.3		
range:	4.6–8.2	3.5–9.0	3.0–4.6		

Mugler (14CY1)[c]

L	W	T	L	W	T
6.8	7.5	4.2	4.7	5.0	4.2
7.0	8.6	4.3	6.3	6.8	4.7
6.0	7.5	4.5	4.8	4.8	3.0
6.5	8.0	4.8	6.0	6.5	3.6
7.0	7.3	4.8	6.8	5.8	3.8
8.0	9.8	4.4	5.2	5.2	3.2
7.0	7.3	4.2	6.0	5.5	4.0
6.8	7.5	4.5	5.0	5.6	4.0
6.8	7.4	4.2	4.2	5.0	2.8
7.0	7.0	4.2	6.4	7.2	3.8
6.0	5.5	3.7	4.3	5.5	3.4
5.7	6.0	3.4	5.3	5.3	3.8
5.8	6.3	4.0	4.7	5.5	3.5
5.0	5.6	3.6	5.6	5.6	3.8
5.7	6.4	4.3	4.7	5.1	3.7
4.2	5.0	2.7	5.0	5.6	4.4
5.4	5.2	3.2	4.5	5.3	3.0
5.2	6.5	3.4	4.4	3.7	3.3
5.5	5.7	4.6	4.8	6.5	3.4
4.4	5.5	4.2	4.6	6.4	4.0
4.5	4.7	4.2	4.0	4.3	2.8
4.5	5.6	4.3	5.8	7.0	3.2
5.4	5.5	3.6	5.0	5.0	3.6
5.5	6.5	4.5	4.5	6.0	3.8
5.0	5.4	4.0	5.0	6.0	3.2

N = 50

	L	W	T	W/L	T/W
=	5.5	6.1	3.8	1.1	.62
S =	.95	1.2	.54		
S^2 =	.89	1.3	.29		
range:	4.0–8.0	3.7–9.8	2.7–4.8		

Ware (from Schultz Collection—no number)[c]

L	W	T	L	W	T
7.5	8.0	5.0	7.3	8.6	5.5
6.8	9.5	4.8	6.5	8.5	4.8
6.8	8.3	4.6	6.8	7.3	4.4
6.2	7.0	3.8	8.0	9.0	5.2
7.0	7.8	5.4	7.6	9.3	4.7
6.7	8.0	4.3	7.0	8.0	4.5
6.8	7.5	4.6	6.8	7.3	4.3
6.5	7.0	4.5	7.3	7.3	4.5
7.0	8.6	5.0	7.4	8.6	4.5
8.0	9.5	4.3	7.0	7.5	4.5
8.0	7.6	5.0	7.0	8.6	5.2
7.0	8.4	5.2	5.7	6.8	4.0
6.7	8.4	4.9			

N = 25

	L	W	T	W/L	T/W
X̄ =	7.0	8.1	4.7	1.2	.58
S =	.54	.78	.42		
S^2 =	.28	.58	.17		
range:	5.7–8.0	6.8–9.5	3.8–5.5		

Kemp (14CY4)[c]

L	W	T	L	W	T
7.0	6.5	3.5	7.7	7.0	4.0

Table A.4. Summary Data on Measurements of *Zea mays* kernels and cupules. (continued)

8.0	8.8	3.6	7.4	8.5	4.5
7.6	7.5	5.1	6.8	7.6	3.7
7.5	8.6	5.0	5.0	6.7	3.8
5.7	6.5	4.8	6.0	6.7	4.3
5.8	6.3	4.0	8.2	7.9	3.7
8.5	8.0	3.4	6.3	7.0	4.0
6.5	6.8	3.5	6.8	7.8	4.6
8.5	8.0	4.2	7.3	8.0	4.1
6.6	7.5	4.0	6.0	7.5	4.1
5.5	6.8	3.7	7.8	8.8	3.8
7.0	8.7	4.8	7.0	7.5	3.3
7.5	7.8	3.8	7.0	7.5	3.4
6.7	7.8	5.2	6.5	7.6	4.0
6.2	8.0	4.8	6.8	7.0	3.8
7.0	8.0	4.0	7.0	7.0	4.2
6.3	7.4	3.7	6.0	6.8	3.7
6.8	7.0	3.9	8.7	8.4	3.2
7.5	10.0	4.4	6.3	7.8	4.3
6.3	7.3	4.6	6.3	7.5	4.0
5.7	6.8	3.7	7.0	6.8	3.2
7.5	9.4	4.2	7.0	8.3	4.5
7.2	7.3	4.8	7.6	8.5	4.2
7.4	9.8	5.0	5.0	7.4	4.5
5.8	7.8	4.8	5.5	8.0	4.4
6.5	7.3	4.5	6.7	7.7	4.2
5.0	6.3	3.8	5.5	6.5	3.3
5.5	6.5	4.0	7.5	6.7	3.6
6.5	8.2	4.9	6.0	6.2	3.4
6.0	7.5	3.8	6.5	7.0	3.2
6.4	7.5	4.7	6.7	6.8	4.0
6.2	8.0	4.5	6.8	8.0	4.6
5.5	6.3	3.3	6.0	6.2	4.0
7.0	7.0	4.4	7.3	9.0	3.7
5.0	6.2	3.4	7.5	8.0	3.8
6.4	7.5	4.3	7.0	8.0	4.7
5.5	6.5	3.8	6.0	6.0	3.5
6.8	7.8	5.0	7.0	8.0	3.6
6.3	7.0	4.4	6.0	6.5	4.0
6.0	7.0	4.7	6.3	7.3	4.0
5.8	7.2	4.0	7.5	8.0	4.0
8.0	8.8	3.6	5.6	6.8	4.0
8.1	8.0	3.4	7.0	8.0	3.8
7.3	8.6	3.4	6.2	7.0	4.5
6.3	7.0	3.6			

N = 89

	L	W	T	W/L	T/W
$\bar{X}$ =	6.7	7.4	4.1	1.1	.55
S =	.84	1.1	.5		
S^2 =	.7	1.2	.3		
range:	5.0–8.7	6.2–10.0	3.2–5.2		

cupule measurements–width

9.6
9.4
8.8
9.7
9.5
7.0
8.0

N = 7
$\bar{X}$ = 8.9
S = 1.0
S^2 = .87

Anderson (14HP5)
cupules: width measurements

9.4	6.0
9.0	9.0
8.9	7.8
8.9	6.9
9.0	3.9
8.4	4.0

N = 12

	W
$\bar{X}$ =	7.6
S =	1.97
S^2 =	3.57
range:	3.9–9.4

Buresh (14SR303)

L	W	T	L	W	T
6.4	7.0	4.0	8.0	8.1	5.0
7.0	8.0	4.5	6.6	6.7	4.0
6.2	7.4	4.0	5.5	7.4	4.7
5.8	7.0	5.1	7.0	7.6	4.8
6.2	8.0	5.6	6.4	7.3	4.6
5.4	7.0	3.7	6.7	7.4	5.5
6.0	6.8	4.0	6.5	6.9	4.5
7.5	8.7	5.9	6.0	6.5	4.1
6.7	8.4	4.4	7.8	7.8	5.5
5.3	7.3	4.8	6.3	6.7	4.4
4.2	6.0	4.2	5.0	5.8	4.7
4.7	6.2	4.0			

N = 23

	L	W	T	W/L	T/W
$\bar{X}$ =	6.2	7.2	4.6	1.2	.63
S =	.94	.74	.6		
S^2 =	.84	.53	.35		
range:	4.2–8.0	5.8–8.7	3.7–5.9		

cupule measurements–width

9.2	8.2
8.7	7.9
8.7	7.7
10.3	9.1
8.9	7.8
8.8	8.0
9.0	7.2
8.6	8.9
8.3	7.4
8.0	7.8
8.6	9.1
8.4	7.7
7.4	6.9
	7.8

N = 27

	W
$\bar{X}$ =	8.3
S =	.75
S^2 =	.54
range:	6.9–10.3

[a] only complete kernels or cupules were measured; no corrections were made for charring.

[b] all measurements are in mm.

[c] represents approximately 10%–25% of the collection.

Table A.5. Summary Data on Measurements of *Phaseolus vulgaris*.

23CL115[a]

L[b]	*W*	*T*	*L*	*W*	*T*
7.1 ×	4.5 ×	3.0	8.6 ×	5.3	
8.9	5.1	4.5	9.5	5.2	
9.4	5.0	3.6	9.9	5.7	
7.8	4.7		10.0	5.1	
8.2	4.5		10.0	5.4	
8.2	5.1		10.2	5.2	
8.6	4.9		10.3	5.4	
8.6	5.1				

N = 15

	L	*W*	*T*
$\bar{X}$ =	8.9	5.1	3.7
S =	.9	.32	.76
S^2 =	.8	.10	.38
range:	7.1–10.3	4.5–5.7	3.0–4.5

23CL109[a]

L	*W*
8.1 ×	4.6
9.0	5.0

N = 2

Coons (23PL16)[a]

L	*W*	*L*	*W*	*T*
8.3 ×	4.3	9.0 ×	5.3 ×	4.2
8.9	5.3	9.0	6.2	3.6
7.3	4.3	7.8	4.5	3.4
9.5	5.6	9.9	5.5	5.0
8.8	4.8	8.2	5.0	
8.5	4.9	8.2	5.2	
9.0	5.4	8.3	4.5	
10.9	6.4	8.3	4.6	
11.1	6.4	8.3	4.8	
8.5	5.0	8.3	5.2	
9.5	5.4	8.4	4.6	
7.0	3.7	8.4	4.7	
7.0	4.7	8.5	5.4	
7.4	4.0	8.6	4.6	
7.4	4.1	8.6	4.7	
7.4	4.4	8.7	4.9	
7.4	4.4	8.8	4.8	
7.8	4.5	8.8	4.9	
7.9	4.9	8.9	4.6	
8.0	4.4	9.1	4.6	
8.0	4.5	9.1	5.0	
8.0	4.5	9.1	5.4	
8.0	4.5	9.2	5.3	
8.1	4.8	9.2	5.3	
8.2	4.5	9.3	5.1	
8.2	4.5	9.4	5.2	
9.4	5.3	9.4	5.4	
9.9	5.2	10.0	5.7	
8.6	5.0	8.5	5.0	
9.0	5.0	9.9	5.4	
9.2	5.0	10.0	5.8	
9.7	5.4	10.0	6.0	
7.8	4.1	7.0	4.1	
8.2	4.8	9.2	5.1	
8.3	5.0	9.7	4.8	
9.1	5.0	9.2	4.9	
10.0	5.4	10.0	5.4	

N = 74

	L	*W*	*T*
$\bar{X}$ =	8.7	5.0	4.1
S =	.87	.53	.27
S^2 =	.75	.72	.38
range:	7.0–11.1	3.7–6.4	3.4–5.0

13ML129

	L	*W*	*T*
Cache 1	10.0 ×	6.0 ×	3.5
	8.3	5.6	
	8.5	4.8	
	9.8	5.3	
	9.3	5.0	
	9.5	5.5	
	10.2	5.8	
	8.9	5.0	
	8.8	4.7	
	10.2	6.1	
323-7L3	11.0	6.0	

N = 11

	L	*W*	*T*
$\bar{X}$ =	9.5 ×	5.4	–
S =	.83	.51	
S^2	.63	24	
range:	8.3–11.0	4.7–6.1	

13ML138

L	*W*	*T*
11.3 ×	7.5	5.2
10.2	4.9	
9.8	7.9	

[a] measurements provided by Leonard Blake, Missouri Botanical Garden, in letters to P. J. O'Brien, 1974 and 1980.

[b] all measurements are in mm for complete specimens only; no corrections made for charring.